WATER RESOURCES IN THE HIMALAYAS
Harvesting, Tradition and Change

Concept's Discovering Himalayan Series

1. People's Participation in Himalayan Eco-System Development: A Plan for Action—*M.L. Dewan*
2. Nut Fruits for the Himalayan—*M.L. Dewan, M.C. Nautiyal and V.K. Shah*
3. Vedic Philosophy for Himalayan Eco-System Development —*M.L. Dewan and B.D. Joshi*
4. People's Participation and Rejuvenation of Himalayan Environment—*M.L. Dewan*
5. Water Resources in the Himalayas : Harvesting, Tradition and Change—*Piyoosh Rautela*

About the Author

Having done masters in Applied Geology from the University of Allahabad, Piyoosh Rautela persued his doctoral work at Wadia Institute of Himalayan Geology, Dehradun. Geological fieldworks in the Himalayan terrain took him close to the groundrealities at the grassroots level and the burning desire to contribute something positively towards improving the quality of life of the masses and deteriorating environment of the Himalayan region brought him in the contact of various voluntary groups working towards this end in the region. He has been actively associated with ENDS Dehradun and TFL/HIMCON Ranichauri for the previous four years and has contributed extensively at the grassroots level in the field of water and watershed management. Fruits of his hard labour are reflected in Lawadhar and Paturi watersheds of Chamba block of Tehri Garhwal.

His environmental pursuits do not deter him from contributing in the scientific field and he has worked extensively on GIS based landslide and environmental hazard assessment, palaeoclimate and structural geology.

He is at present engaged in an independent short-term research assignment of the Department of Science and Technology (Government of India) at Wadia Institute of Himalayan Geology, Dehradun.

WATER RESOURCES IN THE HIMALAYAS
Harvesting, Tradition and Change

PIYOOSH RAUTELA
Wadia Institute of Himalayan Geology
Dehradun

Foreword by

Dr. M.L. Dewan

CONCEPT PUBLISHING COMPANY, NEW DELHI-110059

ISBN 81-7022-804-2

First Published 2000

Published and Printed by

Ashok Kumar Mittal
Concept Publishing Company
A/15-16, Commercial Block, Mohan Garden,
New Delhi-110059 (India)

Phones : 5648039, 5649024

Fax : 091-11-5648053

E mail : publishing@conceptpub.com

FOREWORD

Water is the life-blood of agriculture, it is constituent for economic development of a region and has a special place and significance in the life of the people of the Himalayan region. The Himalayan watersheds are dependent upon atmospheric precipitation for all their water supply. Rains though abundant, are not uniformly distributed and the terrain conditions promote quick runoff of the precipitating water to lower altitudes. This flowing water cannot be brought up without an expense of energy and, therefore, every effort should be made towards conserving it at the higher altitude for as longer a duration as possible. This would maximise recharge of subsurface waters and the water so harvested would be available at easy approach for diverse use.

Floods are a menace and bottleneck to the realisation of the optimum use of resources. These occur practically every year in varying degrees of intensity. It is possible for mankind to reduce but probably not completely stop this devastation. Reduction is possible through water harvesting and watershed management in the headwaters of the river systems.

Droughts have been observed as a setback to the welfare of the people as a flood drought-flood cycle has been causing. Drought brings near famine conditions for the people. Now that drought prone areas have been identified and remedial measures contemplated some worth while action can emerge.

Himalayas, the abode of eternal snow, name given to the vast stupendous mountain ranges on the northern frontiers of India, stretches for about 2,250 km from the west to the east. The northern parts of the ranges are drained by the Indus and the Brahmaputra whilst the southern

Himalayan slopes are drained by the Ganges.

Uttarakhand or Uttranchal as the hill districts of Uttar Pradesh are called, form an important resource potential of Uttar Pradesh and other parts of the Himalayas have important resource potential. Dr. Piyoosh Rautela, a young geologist with experience in water and watershed management at grassroots level has been, as a practical Scientist, working on various aspects of water harvesting, water management and other aspects of water use in the Himalayan Region with special emphasis on Uttarakhand. Dr. Piyoosh Rautela is a talented author with great potentality. His first book *Water Resources in the Himalayas* gives lots of information, material, case studies on water for different uses specially drinking water, for household, for cattle use and for irrigation.

I am sure that all those interested in Natural Resources of Uttarakhand and of Central and Western Himalayan Region will welcome this new book as enriching the knowledge of Water in the Himalayas. I congratulate the author heartily for the good publication.

Ranichauri - 249199
Tehri Garhwal, UP

DR. M.L. DEWAN
Chief Regional Bureau for Asia and
Pacific (Retd.) FAO of UN. and
Advisor, Trees for Life, and
Chairman Himalayan Consortium
For Himalayan Conservation-HIMCON

PREFACE

Water is interwoven into the religio-cultural fabric of the Indian subcontinent. Many important rituals are associated with water and it is rendered a highly ritualistic status. Archeological excavations and historical texts have brought forth magnificent traditions of water harvesting and management. Despite these it is a pity to note that an overwhelmingly large proportion of the populace does not have access to adequate quantities of safe drinking water. In Uttarakhand (hills of Uttar Pradesh), that is the focus of this compilation, the situation is far from satisfactory. As elsewhere, many a practices of the little tradition of the indigenous population revolve around water. In this region the discipline of locating and constructing water-harvesting structures was highly evolved. Much of this traditional knowledge is, however, lost and there is a pressing need to undertake serious and dedicated studies for reviving these. The indigenous people of the region well realise the importance of the vegetal cover in the recharge zone of the springs and thus religio-magical sanctions were aroused to conserve these. The maintenance of the traditional structures was joint responsibility of the community. Weakening of the social fabric, together with growing individualism and dependence upon tap water has taken its toll and a large number of traditional sources of water are reportedly showing depleted yields.

Uttarakhand is the source of many perennial rivers and streams and is gifted with abundant precipitation. Topography together with dwindling vegetal cover, however promotes quick runoff and the hill aquifers are not recharged adequately. This causes soil erosion and reduced resource availability, triggering what is often described as the vicious circle of resource degradation. More than anything else, low water availability in this region is associated with the poor management of the resource, which because of its cheap availability is often not even considered a resource worth planning. If managed in a rationalised manner indigenous resources have the potential of meeting the demand of water. Deteriorating water availability scenario in the region calls for concentrated and dedicated efforts from different quarters; more so from within ourselves.

This compilation is a humble attempt to highlight the deteriorating water availability scenario in the region, with limited field examples. I gratefully thank Drs. Dinesh Sati and R.B.P. Singh of ENDS Dehradun, Mr. Rakesh, Shiv & Gopal Bahuguna of TFL/HIMCON Ranichauri and Mr. Rakesh Goel of IGSSS who cooperated in diverse ways during the preparation of this next. My father, Shri B.S.Rautela, and brother-in-law, Dr. Sher Singh, are thanked for painstakingly collecting important literature/ data. Dr.M.L.Dewan inspired this work and but for his blessings this could not have materialised. Discussions on various aspects of the problem with Dr.R.K.Pant of WIHG were valuable and helped in improving the text. Last but not the least, thanks are due to Putput, my naughty but loving little one, and my wife Manisha who took the pain of correcting the manuscripts and for sparing me to undertake this work, that too with a smile. Partial financial assistance from Indo-German Social Scrvice Society, New Delhi is acknowledged.

13th August 1998 PIYOOSH RAUTELA
Dehradun

CONTENTS

LIST OF PLATES

LIST OF FIGURES

LIST OF TABLES

1

INTRODUCTION

It is water, dubious presence of that paved way for the speculations upon the presence of life on planets other than earth. This fact amply highlights the significance of water in the evolution of life. In the primordial atmosphere devoid of oxygen life evolved around volcanoes and was aided by torrential rains. With the evolution of life photosynthesis evolved and free oxygen first appeared in the earth's atmosphere. Under the effect of solar ultraviolet radiation oxygen was transformed to ozone and the protective layering (the ozone layer) shielding off the harmful ultraviolet rays was formed in the outer atmosphere—the stage was thus set for the appearance of organisms that fed upon the renewable organic matter—the animals. Water thus has the prime credit of making planet earth hospitable to life and has played important role in the survival and evolution of life on this planet.

Water has diverse socioreligious uses and is considered to be most purifying as well as polluting fluid. Water of Ganga, Narmada and Godavari holds a pan-Indian sacredness. Almost all small rivers, springs and water bodies are attributed some degree of holiness and associated with the local pantheon of Hindu gods and goddesses.

Water is a vital natural resource, a must for sustaining any form of life. When the habitation and migration patterns of wild animals are controlled by the availability of water, it is fair to expect the choice of early human settlements being dictated by the availability of water. It is therefore not surprising to note that all the early human civilisations across the globe flourished in the river valleys. The growth of all these is closely linked to creating, managing and exploiting the riverine resources. Lapses in the management often wrote off civilisations from the face of the earth. In the mountainous terrain strategic compulsions often pushed habitations to ridges and the settlement pattern was controlled by the presence of natural seepage and springs. Greater acumen was, therefore, required for selecting the habitation site in the hilly terrain. With increasing population water from natural sources in the hills would have fallen well short of the demand. This exigency, that threatened the survival of the entire community, gave birth to various water harvesting practices. The dictum of natural selection favoured communities with developed sense of water management. With experience, observation and experimentation human populations mastered the art of locating the sites where water could be expected beneath the ground and could be stored for long spells. It is this breakthrough that made possible the evolution and growth of cultures in the mountainous terrain of Himalaya. The fact that every region has its location specific practices testifies that these practices evolved independently. Interestingly the areas that have many natural springs by virtue of favourable structure and lithologic setup, together with high drainage density, have less evolved water-harvesting practices.

The economic importance of water in a country where agriculture is the mainstay can hardly be exaggerated. Indian history is full of examples of indigenous water harvesting systems showing enormous variation in their design, structure and size. Though crude by present day standards, these were nonetheless complex and required enormous skill to construct. Many of these structures continue to meet the needs of the people even today. State management of irrigation was not unknown in the past, but community management was widespread. It was the people's participation in the management of these resources that made it a success.

Irrigation management has several phases, namely, control of water source, delivery of water, and actual application of water to the crops and drainage, each of which involves a number of distinct functions like facility construction., operation and maintenance, water allocation and conflict resolution. Institutions for water control have to be viewed in relation to agro-climatic conditions, technology of water control and agriculture, land tenure and other factors which define the context in which these institutions function and which to some extent condition their characteristics.

Water had both to be managed and exploited for the survival of the human race and, therefore, with the first human settlements around 6,000 years ago began a two-fold struggle with water. On the one hand, people had to protect themselves against floods, on the other; they had to ensure safe water supply for domestic use and irrigation. Hydo-technical installations were thus amongst the earliest technological achievements of mankind.

Art of storing water in cisterns was mastered well before the first cities came into existence. The earliest cisterns were hewn into the rock. Masonry cisterns developed later. From about 2000 BC did mortar began to be used for sealing these cisterns (Garbrecht 1986). By the middle of the first century AD, covered cisterns with storage volumes of upto 75,000 cubic metres were being built and fundamental elements of dam construction were known.

The earliest dams were constructed in Jawa (Jordan) around 3000 BC and in Wadi Garawi (Egypt) around 2600 BC. The first great civilisations, in the valleys of Nile, Tigris, Euphrates, Indus and Huang Ho rivers, flourished only on the basis of their superior water management systems. Both the Indus valley and the Mesopotamian cultures had excellent water supply and sewage systems. The standard of their hydro-technical infrastructure was not equalled even 2,000 years later by the Romans.

The early Hindu texts, written around 800-600 BC, reveal a certain knowledge of hydrological relationships. The Vedic hymns, particularly those in *Rig Veda*, contain many notes on irrigated agriculture, river courses, dykes, water reservoirs, wells and waterlifting structures. The *Chandogya Upanishad* points out: The rivers .... all discharge their waters into the sea. They lead from sea to sea, the clouds rise them to the sky as vapour and release them in the form of rain . . .

This is probably the oldest reference to natural processes within the hydrological cycle. It clearly shows that as early as 1000 BC attempts were being made to interpret and explain recurrent natural phenomenon on the basis of day to day experience.

Growth of flourishing culture in the heart of the world's highest mountain range clearly indicates that the water management practices are imbibed deep in the socio-cultural fabric of the region. Across the length and breadth of this region a variety of innovative practices are encountered. Large population centres of this region that have a long historical presence clearly testify that natural waters had the potential of supporting indigenous population. Over the years something seems to have gone wrong and we have a trail of dying springs... scarcity of water is rampant in this region... and it has become a commodity up for sale openly during summers.

It is in this background that this book is being written to access the availability of water in the mountainous terrain of Uttar Pradesh. However, before focusing upon the region a broad overview of the water harvesting practices is presented on a wider perspective. After discussing some specific case studies a general discussion upon the various reasons for water crisis is presented and in the concluding section some suggestions are put forth for improving the scenario.

WATER HARVESTING THROUGH THE AGES

Water has been harvested in India since antiquity. Ancient texts, inscriptions, local traditions and archaeological remains provide the evidence for this tradition. Rigveda mentioned a lot about the rational and judicious use of water from wells, tanks, ponds and canals. In the *Rigvedic* period, farmers constructed channels for irrigation (Rao 1971). The *Ramayana* allegorically refers to dykes releasing rainwater. *Visva-Vallaha* of Chakrapani describes ways for approximating the depth of subsurface water reserves. While digging if the stone like hard earth is reached and when struck it sounds like a thin slab of stone, then there is sure to be plenty of water beneath.

Excellent traditions of water harvesting are observed in the Indus Valley. French archeologists working in Afghanistan discovered Harrapan settlements in the province of Bactria, near the Russian border. In the Ai-Khanum region they have come across several large canals, each about 25-30 km long, some 30 m wide. Excavations carried out in the area along the hypothetical route of the canals have unearthed pottery confirming the Harrapan affinity. The irrigation technology either emerged locally or was brought by the Harappan settlers migrating northwards from India. Inexplicably, the Harrapans seem to have built an efficient irrigation system in the outer areas of their influence, whereas there is no evidence of such a system in the core area. The lack of evidence may well be attributed to the archeological focus upon prosperous, urban centres.

Well, probably a Harappan invention is a distinguishing feature of the Harappan civilization and has been found at most house sites. It is, however, debatable whether or not Harappans used wells for irrigation. Excavations at Allahdino, a small Harappan settlement near Karachi, have however, revealed a well that may have been used for irrigation. Opinion holds that the diameter of the Harappan wells was deliberately kept small so that the water level could rise higher and could overflow in an artesian fashion. Cultivation of winter crops like wheat and barley suggests that the Harrapans must have had some method of irrigation (Ratnagar 1986). Ditches were probably dug for diverting the water from rivers. At Amri culture site of Kai Buthi, the diversion of spring water to prepared agricultural fields is observed (Possehl & Raval 1989). Nuka in Sind in the Kirthar-Kohistan region, provides evidence of artificial irrigation similar to gabarbands of Baluchistan.

Located on the northwestern end of the Khadir island in the Great Rann of Kutch, Dholavira, a major site of Indus valley civilization, dates back to third millennium BC. This arid area experiences negligible rainfall (annual average 260 mm) with no perennial sources of water in the vicinity. Subterranean water is, by and large, brackish and saline and, therefore, unfit for drinking or even cultivation. The quantity of sweet, potable water is extremely limited and is available only in a few pockets. In order to make maximum use of the available water the inhabitants of Dholavira created several reservoirs for collecting the monsoon runoff flowing in the flanking streams of Manhar and Mansar. Stone *bunds* were raised across these at suitable points in order to divert the flow of water through inlet channels into a series of reservoirs that were dug out in the sloping areas between the inner and outer walls of the Harappan city. Water reservoirs were separated from each other by *bund*-cum-causeways, that facilitated access to different divisions of the city as well. Likewise, a network of drains crisscrossing the citadel was also laid out to collect rainwater.

An organised water pricing system was developed during the period of Chanakya (3rd to 4th century BC) and detailed account of the various practices contained in Arthasastra finds corroboration in inscriptions and archaeological remains. Regarding the activity of the Heads of Departments, Kautilya says: He should build irrigation systems with natural water sources or with water to be brought in from elsewhere. To others who are building these, he should render aid with land, road, trees and implements and also give aid to the building of holy places and parks. If one does not participate in the joint building of an irrigation work, his labourers and bullocks should be made to do his share of work and he should share the expenses but will not receive any benefits from it. The ownership of the fish, ducks, and green vegetables in the irrigation works should go to the king (Kangle 1963).

In other chapter concerning judges he says: In case of damage to the ploughing or seeds in another's field by the use of reservoir, channels or a field under water, they shall pay compensation in accordance with the damage. In case of mutual damage to the fields under water, parks and embankments, the fine shall be double the damage.

A tank on lower level, constructed afterward, shall not flood with water a field watered by a tank on higher level. A tank constructed on higher level shall not prevent the flooding with water of a lower tank, except when its use has ceased for three years. For transgression of that, the punishment shall be the lowest fine for violence and the emptying of the tank (Kangle 1963).

In yet another chapter, Non-observance of Conventions, Kautilya observed: If one obstructs a customary watercourse in use or makes a new one that is not customary, the lowest fine for violence shall be imposed; the same fine will be imposed if one constructs on another's land a dam, a well, a holy place, a sanctuary or a temple. If a person himself or through others mortgages or sells a charitable and existing waterwork, the middle fine for violence is to be imposed, the highest on witnesses, except when it is in ruins and abandoned. In the absence of the owner, villages or persons of a pious disposition should repair these (Kangle 1963).

Two inscriptions engraved on a rock in Junagadh

(Gujarat) provide interesting information about the repair of an embankment that was destroyed during a flood. The first of the Saka ruler Rudradaman dates back to 150-151 BC and records the restoration of lake Sudarsana by Mahakshatrapa Rudradaman. The lake had been constructed during the reign of Chandragupta Maurya (324-300 BC) by his viceroy Pushyagupta and was later improved under the reign of Ashoka (273-232 BC), when irrigation canals from the lake were excavated by the Yavan king Tushaspha. Storing the water of Suvarnasikata and Palasini streams running from Urjayat (Girnar) hill created this lake. The embankment had an earthen core and stone facing on both sides. This time-tested method was in vogue through out the country until cement and concrete started replacing it. Shortly before 150-151 BC, floods of Suvarnasikata, Palasini and other streams caused a breach in the embankment. Rudradaman's minister Suvisakha, a Pahlava (Parthian) carried out the work of restoring the dam. The inscription underscores the fact that people knew how to construct dams, lakes and irrigation systems even in the 4th century BC.

The embankment of the Sudarsana lake burst again as a consequence of excessive rain in 455-456 AD and was repaired under the orders of Chakrapalita in 456 AD. Another inscription at Junagadh, at the time of king Skandagupta (455-467 AD), records the repairs of the embankment by Chakrapalita (Chhabra & Gai 1981). The Sudarsana lake fell into disuse sometimes in the 8th or 9th century AD due to destruction of the embankment and was never restored again.

Vraht Sanhita

By laws regarding the orientation of ponds to store and conserve water efficiently and protecting them from possible damages have been specified in Vraht Sanhita during 550 AD (Jha 1988). The guidelines may be briefly summarised as following:

— A pond laid east to west retains water for a long

time while one from north to south gets spoiled invariably by waves raised by the winds. To stabilise walls, it is necessary to line them with timber or stone and the adjoining soil should be strengthened by stamping and trampling by elephants, horses etc.

— Banks must be shaded with useful trees.
— An outlet for water needs to be made on a side with the passage being laid with stones.

Karyapa *muni* — indegenous genius of our tradition

Karyapa *muni* gives a fairly detailed account of building water reservoirs, site selection and the benefits that would eventually result from them in his book *A Treatise on Agriculture*.

— The king should have a reservoir of water constructed in a most suitable place near villages and towns according to the situation of land to be irrigated. It, therefore, gives due importance to the convenience of the beneficiary population.
— The reservoir should be deep and fathomless like a moat and should either be semicircular, oblong or circular in shape.
— It should have high embankments and be reinforced with strong buttresses and provided with a causeway and channels and sluices for filling and emptying it. Hence, it should be constructed near a hill or on a tableland with a big lake. It should be deep and fitted with floodgates.
— The king should inspect every month particularly in rainy season to ensure safety of the people in case it tends to overflow.
— The reservoir may be fed by a mountain rivulet, a big lake, a forest brook, or a big river according to the situation of the town and villages and should be provided with a suitable contrivance for distribution of water.
— On the banks of the reservoir king should promote

planting of trees such as *Ratan* (*Calamus rotang*), *Banyan* (*Ficus indica*), *Plaksa* (*Ficus infectoria*), *Asvattha* (*Ficus religiosa*), *Khadira* (*Accacia catechu*), *Neem* (*Azadirachta indica*) and *Kadamba* (*Nauclea cadamba*)—these being the trees having hard wood.

— Wells should be dug during the summer season. The king should first ascertain the presence of water by examination of soil and observing the strata of the earth from the directions of the roots of the trees.

Archeological excavations near Allahabad have brought to light an extraordinary example of early Indian hydraulic engineering, dating back to the end of the 1st century BC. The tank is not only the longest of its kind discovered so far—more than 250 m long—but is fed by the waters of Ganga; other contemporary or near contemporary tanks just collected rainwater (Agarwal & Narain 1997).

During the monsoon season, the Ganga swells up by about 7-8 m and spills into a nearby *nullah*. Hence a canal nearly 11 m wide and 5 m deep was dug to take the excess water. The water in the canal first entered a silting chamber, relatively clean water then enters the first tank that was made of bricks. Only clean water could find access to the second tank that constituted the primary source of water supply. The third tank was circular and had an extremely elaborate staircase.

It appears that along its edge stood some shrines in which people offered prayers after taking a ritual bath in the tank. The terracotta sculptures recovered from the debris of this tank include a highly artistic head of a three-eyed Shiva and Kubera. An elaborate waste weir was provided at the end of this tank. It consisted of seven spill channels, a crest and a final exit channel. The excess water was returned to the river, completing the cycle. In case the tank dried up during the summer, the ancient engineers built a series of tanks in the bed of the tank to access the groundwater. Circumstantial evidence suggest that king Dhandeva of Ayodhya under whose jurisdiction Sringaverapura lay probably constructed the structure.

Kalhana's *Rajatarangini* is also replete with information on canals, irrigation channels, embankments, aqueducts, circular dykes, barrages, wells and water wheels. Among the many works mentioned the more significant ones include the huge embankment called the *Guddasetu*, built by Damodara II; controlling the waters of Mahapadma (Wular) lake by a network of canals; diversion of waters of Vitasta and construction of series of water wheels for distribution of water in different villages by Maktapida (725-760 AD). The most important evidence of sophisticated waterworks is about the irrigation works carried out during the reign of Avantivarman of the Utpala dynasty (855-883 AD) by Suyya (Naudou 1971). Having drained off the water of Vitasta, he is said to have controlled it by constructing a stone dam and clearing its bed. Suyya also displaced the confluence of the rivers Sindhu and Vitasta and constructed stone embankments for seven yojans along the Vitasta for damming the Mahapadma lake. The system of irrigation established by Suyya was designed in such a way that everyone was supplied with a fair share of water. Lalitaditya Muktapida is also credited with having supplied villages near Chakradhara (Tsakdhar) with irrigation by constructing a series of *araghatta* (water wheels), that raised the waters of the Vitasta. Perennial water supply was thus ensured through these methods to the arid plateau of Martand, with its magnificent temple, the flourishing township that surrounded it, and his capital Parihaspura (that now suffers water scarcity).

Feroz Shah Tuglaq (1351-1388 AD) stands out for his remarkable public works, more particularly for building five irrigation canals, several dams across rivers, reservoirs for irrigation and repairs to earlier works (Banerjee 1967). After finding the city of Hisar Fefozah, Feroz Shah Tuglaq, perceiving scarcity of water, resolved to bring a supply there. He accordingly conducted two streams (*jui*) into the city from two rivers; one from Yamuna and other from Sutlej. That from the Yamuna was called *Rajiwah* and the other *Alaghkhani*. Both these streams were conducted through the city of Karnal and after wending through a length of about 80 *kos* (about 258 km) discharged their waters by one channel into the town (Banerjee 1967). Shams-e-Siraj Afif presents a vivid account of how numerous waterworks were brought into Fatehabad and Hisar Ferozah and how many towns and villages, were brought under cultivation in his Tarikh-e-Feroz Shahi (Dowson 1953).

Facts

May be the common people do not care but the available water resources are limited. 97.5 per cent of the total available water on our earth is salty and the other 1.76 per cent is locked in the permanent ice caps. The supply of fresh water on the planet is severely limited and fixed. Alteration to this quantity is not possible.

Pollution renders water unfit for human use and the level of pollution is rising alarmingly all over the globe. 4 million children die prematurely every year due to the various water borne diseases. Despite comemorating a decade of availing pure potable water under the ageis of UNO, 31 per cent of the developing population do not have access to safe drinking water.

Feroz Shah Tuglaq also built embankments for storage of water in the close vicinity of his hunting lodges, Malcha Mahal, Bhuli Bhatiyari-ka-Mahal and Kushk Mahal (Sharma 1974). Shams-e-Siraj mentions several other *bunds*, Fath Khan, Malja, Mahapalpur, Shukr Khan, Salaura, Wazirabad, and other similar strong and substained *bunds* (Dowson 1953). Feroz Shah also repaired the Surajkund, constructed in the 10th century by Surajpal of the Tomar dynasty (Sharma 1974). Close to Surajkund is Anangpur dam with sluice opening ascribed to Anangpal, also of Tomar dynasty. This dam is still in use.

India had a long and outstanding tradition of money economy that probably made it the richest country in the world throughout the first millennium of the Christian era. Indian villages functioned like republics and played an important role in managing their own resources. With the coming of British, all major social and economic institutions, including those relating to the village natural resource management declined. British administrators destroyed the financial resource base of the Indian villages and their internal capacity to manage their natural resources. Due to increased land revenue the village community could no longer raise enough internal resources to manage its irrigation structures, thus, agriculture outputs and the income declined to a point that India became a nation chronically affected by famines and destitution.

Throughout the first millennium of the Christian era, India's trade was widespread and Indian merchants controlled many foreign markets (Nehru 1946). The output of Indian peasants was higher in 1700 AD than in 1900 AD (Habib 1969). The uncultivated land then offered peasants extra space to support more grazing. This in turn, meant more drought animals for ploughing, and more dairy products. With the decline of village republics state patronage was required for maintaining the traditional systems. The alien rulers hardly paid any attention to this and the centuries old system crumbled.

After 1857, the country came directly under the British Crown. The subsequent years saw the decline of all the major social and economic institutions in India, including those relating to agriculture and water management. It was the surplus generated in the rural areas that made India one of the world's richest and most urbanised countries when the British first landed on its shores. Due to its monsoonal weather, good water management was critical for good urban and rural development. It was a tradition that India had built up since ancient times.

The land around each Indian village had been transformed over the centuries into a complex ecosystem of croplands, grazing lands, and forest and tree lands, thus constituting an interactive, multi-component biological system that responded not only to the region's sharp seasonal rhythms but also reduced risk by keeping the social and economic impact of rainfall variations down to a minimum. Due to the seasonal nature of rains, numerous types of water harvesting practices developed to meet the water related needs round the year.

The locally developed technologies were supported by an elaborate system of property rights and religious practices. Not only the cow, but the grazing lands had a religious significance. Certain forest land was set aside as sacred groves. The rivers, springs and tanks themselves and their catchments acquired religious importance and would not be polluted.

Agriculture in the mountainous regions needed carefully fine tuned systems of water supply, distribution and management. And so did the towns and cities of the region that could not develop or survive without good water management.

The British in their desire to rule, administer and exploit, destroyed the water management system built by the rural communities of India over centuries. Their revenue system that denied all the customary rights impoverished the peasantry. With the introduction of private property rights in land, village headman and the state appropriated for themselves the ultimate ownership of the soil. This led to the sale of the peasant's land for revenue arrears and the peasants had to surrender their traditional duties to the bureaucracy created by the colonial overlords. This reduced people's participation in the management of their resources. Land steadily became a commodity, and changed hands like any other tradable good. Traditional Indian arts, crafts, and industry and the stock of traditional knowledge, deteriorated without state patronage and the rural gentry, began to loose out. Not surprisingly, there was a general decline throughout the country in indigenous systems of water harvesting and management, with the advent of the British rule. Sir William Willcocks, a British irrigation expert, invited by the British authorities in the 1920s to advise them on why India was suffering from repeated famines, simply ended up advising them on the ingenuity of Bengal's flood irrigation system and his key lesson was that the British would do best to learn from the natives themselves. But this, of course, was too much for the arrogant colonialists to take. But the worst to befall the tradition of water harvesting was the arrogance of the post-independence Indian political leadership and the irrigation bureaucracy, well heeled by then in the ways of the departing white masters, so much so that the brown sahibs gave these systems the biggest slap in their face. Rather than being proud of the rich heritage they did not even care to understand their own indigenous systems, remaining equally arrogant as the British. No effort was made to revive these systems or build new systems in a way that complemented the traditional systems. As a result the traditional systems went into a tail spin. Only in times of scarcity, the Indian media and leadership tend to remember these systems that are calmly forgotten thereafter. It is clear that the people working as noble individuals and as community, have always had a very important role in constructing and managing the traditional water harvesting systems. Local communities, as in the past, could have rebuilt and maintained these systems. The land revenue system introduced by the British, however, deprived them of resources to do so. Unfortunately, there are still very few studies on how water harvesting systems were traditionally managed by the people.

Today, unfortunately, this entire heritage, cultural and technological, lies in tatters. And no agency can be blamed more for this than we ourselves who have allowed our heritage to be lost into oblivion. Technological know-how backing the site selection procedure and estimation of water reserves could not be traced anywhere despite through studies. The present day population attributes it to mere chance but then we should have got a trail of structures that could not fetch water.

3

SCENARIO IN UTTARAKHAND

As social intelligence grew, people realised that the human society cannot grow without extending the bounties of monsoon water from wet months to the dry months. And thus slowly grew the extraordinary traditions of water harvesting in myriad forms. Wherever there were streams, people developed techniques to divert its water, with the help of simple engineering structures, into artificial channels that would feed the agricultural fields. In areas of good groundwater regime, people harvested water with the help of specially designed structures (*naulas*), local variations of which are encountered throughout the terrain. When no options were available people learnt to rely on rainwater for survival. For irrigation purposes they would build rainfed tanks to provide irrigation water for a downstream command area or for cultivating the tank bed itself. People thus did whatever they could—and they did a bewildering variety of things—within the constrains that the local ecology imposed upon them.

Lying between rivers Kali and Tons, Uttarakhand region forms the watershed of the rivers Ganga and Yamuna and is spread over 51,000 sq km. These mountains support about six million people with a population density of 116 people per sq km. Terrace agriculture is generally practiced on the slopes, and paddy cultivation is resorted to in the valleys and duns; the wide valleys that separate the main Himalayan ranges from the Siwaliks. Average landholdings of about 75-80 per cent of the population is about 0.8 ha and irrigation is not available to more than 10 per cent of the cultivated area. The best crops are grown at an altitude of about 915-1,525 m.

There has been an extensive system of water harvesting in this region. Farmers have had a major tradition of building canals roughly aligned with contours to draw water from hill streams or springs. On an average rainfall in the region is around 1,946 mm, about 85 per cent of which falls during the monsoon months of July to September. About 40 per cent of this is lost as runoff. The irrigation is undertaken through *guhls*. The agriculture is largely dependent upon rainfall and the availability of water determines the cropping pattern (Pant 1989).

Despite plenty of rains numerous villages and towns of the region face an acute drinking water shortage. Yet, it is amazing to note that the people today remember traditional water harvesting systems only when there is an acute water emergency. Human society must learn to live with their environment. Regardless of the quantity of water nature provides a society, if it cannot learn to respect and store the water it receives, no quantity of water will ever be enough to meet even the simplest needs.

Many villages in the region have rainfed earthen tanks (*khals* or *chals*), large enough to serve as watering ponds for two months for the village cattle. There are numerous *naulas,* which are lined percolation storage tanks that collect near subterranean flow. In heavily populated centres, *naulas* were the sole source of drinking water before piped water supply came into being. Even today these supplement potable water.

The enactment of the Kumaon and Uttarakhand Zamindari Abolition Act of 1950 confirmed that the ownership of a private well vested in the owner of the land in which it was located. Under the Rule 41(vi) of the Act, it was stated that tanks, ponds, ferries, water channels belonging to the state shall be managed by the Gram Sabha or any other local authority established. Against this background the state government passed Kumaon and Garhwal Water (Collection, Conservation & Distribution) Act of 1975. Section 3 of the law terminated the current and customary rights of individuals and village communities. Under Section 4 of the Act, the state took over the power to frame rules for collection, conservation and distribution of water and control of water sources. It, however, provided for giving priority to such village communities from whom the powers had been withdrawn. No one was allowed, without receiving prior written permission of the subdivisional magistrate, to create an irrigation system. This piece of law undermined people's participation in the construction, management and maintenance of water harvesting structures.

Traditional water harvesting systems of Uttarakhand

Systems for agriculture	Systems for drinking water and other household needs
(a) Diversion channels leading directly to agricultural fields (e.g. *Kuhls*)	(a) Natural springs were often harvested
(b) Occasionally, the channels first lead to a storage structure so that the water can be used in the subsequent dry period, too	(b) Rainwater harvesting from rooftops
	(c) Groundwater harvesting structures (e.g. *Naula*)

Since antiquity, the people of Uttarakhand have drawn water from small water harvesting structures (*naula*) and from natural springs (*dhara*). *Naula* is method of groundwater harvesting whereby the water is collected by making a stone wall across a groundwater stream. Villagers traditionally held *naulas* in great reverence and rituals observed while constructing a *naula* were similar to that of a temple. For treatment and purification of *naula* water, leaves of medicinal plants and fruits like *anwla* were added periodically. Big shady trees were planted near *naulas* to reduce evaporation. It was a custom of the local people to worship *naulas* and trees. This had a two fold effect—it taught villagers to keep *naula* clean and to conserve water, besides protecting the vegetal cover in the recharge zones of the same. The technology of *naula* construction is very old. Despite thorough probe no information could be gathered on the method employed for site selection and estimating the expected discharge.

Various anthropogenic reasons have, however, caused majority of the *naulas* to dry up and have pushed the rest to the verge of extinction. Most important cause of this is overwhelming dependence upon piped water, prolonged ignorance of the traditional structures, and breakdown of the social mechanism for maintaining these structures.

Ironically every third village of the region is affected by an acute water shortage. The State Government has already spent thousands of crores on drinking water supply in the hilly region. It is, however, possible to rejuvenate the *naula* that can reduce the burden of the public exchequer. Temporary stone *bunds*, and check dams can be erected at suitable sites to enhance recharge of the *naulas*.

Naulas are a common water harvesting structures in Kumaon but this technology is less developed in Garhwal that is gifted with many perennial rivers and large number of *dharas*.

Maletha *gul* of Tehri Garhwal

As the folklore narrates, this *gul* in Maletha village of Tehri was built by Madho Singh around 1585, during the reign of Mahipal Shah, the ruler of Tehri state. Stung by sarcastic remarks of his beloved, this *gul* with a 100 m long tunnel about 350 m from the diversion head, became the sole purpose of Madho Singh's life. Water, however, could not flow beyond the tunnel. Then Madho Singh, on the orders of the village goddess sacrificed his son Gaje Singh, after which the water started to flow. As the mark of eternal gratitude, the villagers offer the first portion of the new grain to the memory of Gaje Singh, even today. The story is immortalised in the folklores of Garhwal that is narrated even to this date.

Even though this is now a state owned *gul*, the system is run by active participation of the villagers. Each year, before the diversion structure is built a pancake using 50 kg each of both wheat flour and jaggery, is baked at the site of the diversion head and offered to the goddess for whom a small temple has also been erected. Pieces of this offering are then distributed amongst the households, at least one member of which has to be present for building the weir. It is only after completing this ritual that parties are sent to the forest to collect a suitable tree (to be used in the diversion structure) and bring it to the site of diversion weir (Pande 1995).

Water related needs of the hilly people were for generations being met by the local sources; streams, springs and rivers. In most areas these were the springs that traditionally fulfilled the demands of large population. These have mostly been abandoned and the dependence upon tap water has become total. The discharge of the springs in this region is reported to be depleting over the years; many have dried up while many more have become seasonal. The whole of Uttarakhand today suffers from acute water crisis. Even in the urban centres, like Dehradun, Almora, Pithoragarh, Pauri, Chamba, Mussoorie, Lansdowne, the situation is far from satisfactory. In most areas it is the stream water that is being diverted to meet the demand of the people and in summers when the demand is at its peak the discharge of these sources is generally low. Even in the river valleys, such as Bageshwar, Uttarkashi, Srinagar, the availability of water is seriously marred during the rainy season when the stream discharge is laden with volumes of suspended load. The situation in the remote villages is worse and deprived of an easy access to drinking water the people in the hills have to walk for tiringly long distances to collect water. The distance of the water sources has increased considerably in the previous decade and in the changed scenario, quantum of burden upon our womenfolk has increased tremendously. It thus takes a major portion of the productive time of the womenfolk (Plate 1). They are thus deprived from devoting their energies towards other productive pursuits that could aid in improving their economic status and quality of life.

It is really an irony that even after five decades of

independence an overwhelmingly large population of Uttarakhand does not have easy access to adequate quantities of safe drinking water. Not that efforts have not been made in this direction by the governace of free India; massive schemes have been erected, millions of rupees have been pumped in, but the situation is far from satisfactory and is deteriorating with each summer that passes by. This region has become the graveyard of many large, medium and small schemes. It seems that something is really wrong with the planning and implementation of these schemes and unless it is set right there is little guarantee of the future ones fulfilling the desired goals; it is high time that a serious introspection is done towards locating this lacuna. Most schemes for the area are finalised in the state capital, far off from the place by a bunch of decision makers who have scant knowledge of the physiographic condition of the hills. Moreover, these people don't have any social obligations towards Uttarakhand and their decisions are guided by their own vested interests.

If there are hamlets around in the hills there must have been water at a point of time when people inhabited these. Now the reality is quite different; there are agricultural fields, there are cattle heads, there are settlements and of course people but water is nowhere in the vicinity. Many of the productive springs have become seasonal and even dried up, the vegetation pattern is becoming xerophytic. One with slight common sense can tell that the water retention capability of the area has been hampered.

Most of the state-sponsored schemes are target driven and there is no provision for treating the recharge areas, even detailed hydrological studies are not undertaken before clearing these projects. Its sole aim is to make our society dependent upon external inputs by indiscriminate tapping of the water at higher reaches. It adversely affects the discharge of the downstream springs, vegetation and of course the life of the people. The acumen of the local people needs due recognition and ones from amongst the masses, those who can identify themselves with the joys and sufferings of the masses and understand their culture should be given a chance to lead the mantle of these developmental activities. Many NGOs in the region are undertaking pioneering work in this direction with the involvement of the local communities. Hope lies with the strengthening of the NGO movement and association of dedicated persons with it, along with the full participation of the beneficiary populations.

4

OVERVIEW OF THE LITTLE TRADITION OF THE REGION

Religion is commonly perceived as pertaining exclusively to the spiritual domain while science and technology are viewed as systems pertaining exclusively to the material domain. Issues of material development and progress have, therefore, been left to the complete charge of scientific experts while religious beliefs have been left to guide the inner spiritual development of private individuals. An in-depth and unbiased investigation of the religious and cultural heritage reveals that these have, over centuries, provided knowledge of nature and rules of sustainable natural resource utilisation that has enabled ancient societies to survive materially.

A reappraisal and reinvestigation of the relevance of these systems as guides to man's relationship with the environment are important from the point of view of material survival. Destruction of resources and the erosion of life support systems leads to destruction of livelihoods and thus to migration. Religious traditions have a dual role in the search for peace in contemporary times. At the level of resource use these can provide the value framework which makes prudent and hence, sustainable use of resources possible and thus creates the material conditions of peace. At the level of social conflicts, these can prevent the social consequences of resource degeneration from taking an ugly form, by reminding the people that religion teaches peaceful co-existence.

Ours is an ancient civilisation with continuity

extending over millennia. The earliest recorded thought found in *Rig Veda* begins with a vivid description of five elements (*panchatatva*)—earth, water, energy, air and space—that constitute our universe and provide the life support for man and animal and the base for all forms of vegetation and all forms of human activity. These elements are defined as the manifestation of the absolute and man is ordained to conserve these elements for his eternal prosperity by scrupulously performing his duties prescribed in this regard in the various religious texts.

The *ashrams* of the ancient sages represent systems of ecological management marked by the most rational husbandry of natural resources, partnership between man and animal, sound balance between material satisfaction and spiritual and cultural pursuits and community life that ensured health, happiness and harmony for all the constituents.

The *Yagnyavalka Smriti* spells out man's obligations towards protection of the plant and animal kingdom for his own survival and prosperity. Shankara, Madhwa and Ramanuja—all the three schools of philosophy are in unison in stressing the importance of conservation of natural resources and their judicious use by man for satisfying his material needs.

No religion justifies the destructive attitude modern man has developed towards his environment. All urge man to treat nature as a god given gift. Hinduism expresses the firm belief that the natural environment in which man is placed is created by the lord. *Akasa* (space) is the first creation of the lord, and the Brahman in the form of *akasa* created *marut* or *vayu* (air) and *vayu* in that form created *tej* or energy. The other element *apa* (water) and *kspiti* (earth) are further creations. It is affirmed that the lord alone, residing in these various forms of *akasa, vayu, tej, apa* and *kspiti,* created the further entities. The supreme is immanent in every one of these and these are not mere inert and independent entities.

Of the four purusharthas; *dharma, artha, kama* and *moksha; dharma* denotes righteousness or the proper utilisation of the resources to fulfil or satisfy the needs and desires of the humans. If the man goes on increasing his own need than he requires more and more resources to fulfill his needs. *Dharma* etymologically means the stabiliser, the society becomes stable only on account of this. Over-utilisation of resources to fulfil the innumerable needs and wants is prohibited by *dharma.* In fact, *dharma* restricts utilisation of the resources only to satisfy the basic needs of man.

The water of most of the rivers is considered holy. Every *Srivaishanava* had to chant several *mantras* and *slokas* while taking bath. Holy water of celestial Ganga was to be kept in the house of every *Srivaishanava* so that it could be drunk at the time of leaving this mortal world. Therefore, elementary rules against pollution of water were already formulated, concerning in essence simple rules for hygiene.

The ever increasing ecological crisis in all parts of the globe are an indicator of the inadequacy of contemporary scientific beliefs in providing the basis for sustaining the present forms of economic development and natural resource utilisation. The rich philosophical and cultural heritage in the Indian tradition did not provide a set of religious beliefs that can be called totally irrational and devoid of any material basis. Through religion many social customs that were created, only ensured built in ecological restraint. Though not explicitly, but implicitly, the social and religious customs provided the basis for a sustainable and ecological use of natural resources. The heritage of India, undoubtedly holds in store important sources to respond to the ecological crisis emerging in modern India.

In Uttarakhand the indigenous populations had struck a fine tuned balance with nature. These people never overexploited their natural environment. The sons of the soil well understand the fragile ecological relationship and have been for centuries living in perfect harmony with nature. Through centuries of experience these people had learnt to thrive on minimal available resources. Many socioreligious sanctions were woven around forest and water in order to conserve these resources. Though Hinduised little tradition of the people of Uttarakhand is poles apart from the orthodox Hinduism. *Jagars,* spirit and nature worship together with animal sacrifice (*bali*) are imbibed deep into the little tradition of the region and the cultural *digvijay* of Shankaracharya failed to obliterate these. The religious beliefs of the people exhibit a happy amalgamation of little and great traditions.

Woven around the beliefs of little tradition of the region are many a sanctions that protected the natural environment of the region for centuries. The intricate relationship between water and the faunal cover was well understood and the protected forest cover was, therefore, maintained on the higher reaches (the recharge zones of the natural springs) and local deities (*devatas*) were often installed at such places. These were thus looked upon with reverence and special attention was paid to protect their sanctity. Leave aside lopping; even shoes were forbidden in these sanctuaries to keep them pollution free. This was an evolved form of social fencing that was once deep rooted in the hearts of the masses. Religious sanctions associated with these ensured that the God fearing people abide by the norms so set. With changing times these sanctions are, however, being forgotten, and consequently the once thriving sacred groves are degenerating fast.

The places where water was collected were considered auspicious. Many important rituals were associated with these places and deities were often installed at these places (*naulas* and *dharas*). Even today the newly weds are taken to the household *naula* to get the blessings of prosperity. The newly wedded female is thus sought to be as productive as the perennial spring; for in the earlier times it was the fertility of the women that was prized the most for the continuation of the lineage. Under the changed scenario, though the people are finding it hard to forget the age-old practice, the tap is replacing traditional water harvesting structure and important ritualistic ceremonies are done at the location of the tap. This is an example of accommodation, and evolution of the practice through changing times. The cleanliness of the ritualistically important water harvesting structures was also paid due attention and was a social responsibility. In Himachal, Almora, Nagthat, and Tehri Garhwal where detailed fieldwork was undertaken many religious sanctions associated with the traditional water harvesting structures were observed.

With the onset of consummerism and reducing dependence upon the society, the sacred feeling that was the result of the nature-man-spirit complex developed after a close interaction of centuries has been disrupted. But the appeal to the indigenous communities can effectively be made through the use of these motifs and such an attempt is bound to enhance people's participation in eco-restoration work that would have a far-reaching impact upon the environment of this region.

5

WATER HARVESTING TRADITIONS OF CHANGAR

Situated in Khundia tehsil of Kangara district (31°40'—32°25' N & 75°35'—77°5' E) in Himachal Pradesh, in the vicinity of the famous Hindu shrine of *Jwalaji* at Jwalamukhi, this region is locally referred to as *Changar* that may be defined as a part of Kangara district in Beas basin that lacks irrigation, is hilly and where agriculture is still backward.

Changar falls in the sub-tropical climatic zone. The summers over here are hot and dry, winters cool and moist, while the rains are heavy. Though the average annual rainfall in this region is around 1200 mm, it is not well distributed around the year and majority of it is confined to the rainy season. These extremes in climate often enhance the rates of erosion and evaporation in this region. The soil profile in *Changar* is generally shallow and erosion by the forces of nature often exposes the hard bedrock (sandstone), locally referred to as *sappar*. The rocks of the region generally fall under the Siwalik Group and the rock types encountered are sandstone and pebbly sandstone with minor clays, mudstone and siltstone. The sandstone is generally coarse grained, gray to reddish gray in colour, has good porosity and is capable of holding vast volumes of water. Harder varieties of sandstone are also encountered along with minor clays. Both confined and unconfined aquifers are encountered in this region.

The availability of water has always affected the settlement pattern. In *Changar*, the northern slopes are well habitated while the southern slopes are sparsely populated. When people first habitated this region some sources of water would have existed in the vicinity. These sources were subsequently tapped and developed, while the others were designed through long experimentation. The tools, implements and the techniques used by these people for locating and estimating water reserves are lost with the passage of time but the structures developed by them stand testimony of a well developed hydrology discipline that guided them to set forth the roots of the sound traditional system of water harvesting. May be through intuition, but those people could very well locate the sites for these structures. The efficacy of their deep knowledge is testified by the very fact that even after the introduction of tap water through the Department of Irrigation and Public Health (IPH) a major part of the drinking water requirements are still fulfilled by these age old structures. No doubt, in many circumstances these are unable to feed the entire populace; specially during the summers. But then, this can be attributed to the neglect of these structures over the years and indiscriminate tapping of resources by the IPH at higher locations that has ultimately resulted in the lower yields of the downstream structures. Though neglected traditional water harvesting structures are equally relevant in the present scenario. Integration of these into the mainstream has the potential of providing water to the masses in a cost effective manner. It is indeed an irony that even after a trail of failures in providing water to the masses the water bureaucracy is unable to appreciate the potential of the traditional genius.

Everywhere the people develop some innovative practices that can rightly be termed as the tools of adaptation. In *Changar* the topography is rugged, the northern slopes thought gentle are largely *sappar*, the southern slopes are steeper with some grasslands and vegetal cover. The climate, durinal temperature fluctuations and topography of the region favour high rates of erosion and evaporation. Through centuries of experimentation and hard work the people of *Changar* have devised methods for water harvesting for various uses; the ones that are well suited to the ground realities of the area.

Traditional Genius of *Changarites*

Drinking and Household Purposes

Bawari

It is the main source of drinking water in *Changar* (Plate 2). This is a generally a square structure with flights of stairs on all the sides. There is a natural seepage of water within the structure. The source is generally roofed and a wall is erected around it to ward off the animals and also to prevent used water from polluting the water source. Smaller and larger variations of *bawari* are called *bouru* and *naurh*. Provision is made for washing of clothes in the vicinity and due care is taken to ensure that the used water does not pollute the source.

Peepal tree regarded as purifying and some sorts of worship stones are often observed to be associated with *bawari* and is an example of nature-man-spirit complex. Those worshipped are *nag devata* and the ancestors. There exists architectural and ritualistic similarity in similar structures found in different parts of Himalaya (*naula* of Kumaun).

With the passage of time the social system of maintenance of these structures has been disrupted and consequently many have silted (Plate 2), covered with vegetation (Plate 3) and consequently the discharge has gone down considerably.

The traditional method of locating the *bawari* site is lost and despite hard probe nothing can be commented upon this aspect. The people, however, attribute it to the guess work and experience of the local masons that dictated the selection procedure.

The *bawaris* studied in the field were observed to have noticable discharge even during the mid of May. The data on the discharge of some of the *bawaris* is given in Table 5.1 that indicates that on an average a daily discharge of 14,400 liters can be expected from each *bawari*. This data raises vast hopes towards making the community self reliant, considering the number of *bawaris* each village of the region does have (Table 5.2).

TABLE 5.1: **Data on the discharge of some *bawaris* measured in the field**

Location	*Discharge per minute (in litres)*	*Discharge per day (in litres)*
Chillag	3	4320
Bal	4	5760
Salihar	15	21600
Gal	18	25920

Khatri

With depleting water resources rain water harvesting is gaining ground everywhere at present. Rain water harvesting structures are however not very common in the Himalayan terrain. Generally, fractured and foliated strata of this region do not favour this practice. *Changar,* however, stands out as an exception and rainwater harvesting is extensively practiced in this region. Evaporation losses in *Changar* type of climate are extremely high and it is a must to devise some means for countering these for successful storage of water. The best-suited answer to this problem could be subsurface storage and this is what is widely practiced in *Changar*.

Khatri is the traditional rain water harvesting structure of *Changar* and is carved out in the massive, hard sandstone (Plate 4), but those constructed in pebbly sandstone are also observed. The rock type chosen for hosting these structures are free of structural weaknesses and is less permeable. This helps in water storage for prolonged periods. It is a subsurface structure with a small opening on the surface. Stairs are carved out in bedrock and these lead to the

PLATE 1: Quest for water does not end with advancing age and takes toll of the productive time of the females in the hills

PLATE 2: *Bawari;* the traditional water harvesting structure of Changar. Prolonged neglect has however led to the siltation of many in the region

PLATE 3: Breakdown of social mechanism for the maintenance of bawaris has led to the growth of wild vegetation around a bawari whose roof has been destroyed over the years

PLATE 4: *Khatri*, the rainwater harvesting structure in sandstone

underground storage space. The stairs are generally ten in number and each is one foot high. Water percolates to *khatri* by seepage in the overlying strata during the rains and is subsequently used for household purposes. Most *khatris* observed were designed to collect water from seepage in order to retard siltation and also to ensure water quality. Surface runoff is not allowed to enter these structures by erecting diversion channels above the main opening and by keeping the main opening above the ground level. *Khatri* water is not generally used for drinking. These structures are also called *talai* and *kufri*.

On an average the *khatri* has a storage capacity of around 30,000 to 1,00,000 litre. The evaporation losses from these are minimal and the water is stored for long duration and is replenished by occasional rains. These serve the household requirements of the masses and are the lifeline of *Changar*.

The number of the drinking water sources in the villages whose data could be accessed are given in Table 5.2. From this data it is clear that considering the average size of the family to be around 7, the average population of the villages is around 126. Considering the average water requirements to be around 40 litre, the daily water requirements of the villages fall in the tune of 5,040 litre, that can be made available easily from the surrounding *bawaris* and *khatris*.

Irrigation Purposes

Chappri

These are small depressions in the hill side generally at the slope break. Water collects in these during the rains and is used for feeding the animals and for irrigation later on. Besides providing water, *chappri* enhances recharge of ground-water, retards quick runoff and thus erosion. Disruption of *chappri* has led to the depletion in the yields of the downstream *bawaris*. These structures are also observed in Garhwal Himalaya, where these are termed *chaal* or *khal*. These are generally clay lined; *pukki chappri* is also observed

TABLE 5.2: **Account of the water sources in some villages of *Changar***

Village	*Number of Families*	*Bawaris*	*Khatris*	*Taps*
Fareha	14	1	3	
Rihari	15	4	2	3
Bamdiara	20	1	2	1
Thatti	14			1
Bahalan Kurd	22	1	5	3
Bahalan Kala	5	1	2	2
Lihas	12	1		
Daklehad	24	3	1	
Kwali	27	2	4	3
Karal	13	1		
Rihari	9	5		1
Fulwari	8	2	3	2
Jol	9		4	
Upper Mora	8	3		2
Tiala	16		4	4
Badog Lahad	26	2	4	2
Sudra Lahad	20	1		2
Silah	13	2		3
Ambada	67	4		4
Average per village	18	1.8~2	1.8~2	1.7~2

Source: ERA, Khundia.

in *sappar* (Plate 5). While enhancing the storage capacity of the *chappri* care should be taken to ensure that the clays at the base of these do not get disturbed which may lead to leakage of water.

Khu

At many places in the *Changar* the water table is relatively shallow. At these places the subsurface water is utilised for irrigational purposes by digging out wells at specific locations. Water is generally lifted from these wells by load and bucket method. *Khu* are generally stone lined, while the undeveloped wells are called *dungali* (Plate 6). These are traditional dugwells of *Changar*. The traditional genius utilised for locating the depth of the water table could not be traced. In some incidents the water was reportedly struck accidentally and these sites were later developed into *khu* or *dungali*.

Tank

Tanks are used for harvesting the rainwater and also for storing the diverted water at appropriate locations so that it could be convenientiy used for irrigation. These are generally stone and cement lined. At present many farmers are observed to use these for storing the water lifted from the nearby channels.

Bundhs

These are generally constructed at the slopebreak in seasonal *nalas* for storing the rainwater for the lean periods. The storage capacity of these structures soon diminishes due to siltation and lack of proper management strategy. Construction of a series of *bundhs* along the slope could store huge amount of water besides enhancing the recharge of the springs in the downstream area.

Some More Options

In the changed scenario the village community is exposed to many new options that were not known before. There are taps that provide water but the social acceptability of taps in the rural areas is relatively low. People that have access to tap water were observed to collect drinking water from the nearby *bawari* whose water is considered pure. With the growth of technology handpumps are being installed almost everywhere. The water from the handpump is

PLATE 5: Pukki Chappri in sappar near Thill

LATE 6: Dungali the dugwell of Changar at Koke-Baggi with load and bucket arrangement to lift the water

considered pure and is used for drinking purposes. LDPE tanks are also being used to harvest water during the rains. The shallow soil profile and the rocky substratum however create problems in the construction of these tanks. Due care should be taken to reduce the evaporation rates from the tanks. Locally available bamboo can be used for this purpose.

On an average there are two *bawaris* and *khatris* in every village. Daily drinking water requirements of the village come around 5040 litre. The two *bawaris* alone can supply 28800 litre of water daily. Average rainfall in the region is 1200 mm and there are around 18 houses in every village. On an average the covered area of each house is estimated around 40 sq m. If the rainwater on the roofs is harvested with the help of split bamboo pipes and collected it augments the daily supply of water by another 2367 litre. These data show that the villages of *Changar* have a substantial daily surplus of water that comes to 26000 liters. There would still be huge surplus water for routine household needs even after satisfying the animal population of the village. We have not yet added to this surplus the water from other sources like *khatri, khu* and the others. This excess water can be used for irrigating the nearby fields. This data shows that the ancestors in *Changar* devised a technique that can sustain the entire population without any problem. The problems have been created by the enhanced dependence upon the tap water, break down of the social mechanism for the maintenance of the traditional structures and failure to make innovations in the management of the available resource. For a sustainable solution of the water related problems of *Changar* in a cost effective manner the traditional water harvesting structures have to be incorporated in the routine preview of IPH.

6

ALMORA
What Happened to the 360 Naulas?

Situated on a saddle shaped ridge between Kosi and Suyal rivers Almora town is spread over an area of 9.03 sq km and the elevation varies from 1500 m to 1800 m above mean sea level. The average annual rainfall in the town is around 1040 mm and more than 40 per cent of it is received in the rainy season. This township was founded by Chand king Balo Kalyan Chand in 1563 AD and the capital of the Chand kingdom was subsequently transferred over here from Champawat. Gifted with numerous natural water sources, this stratagically important region was until then under the Katyuri kings.

Facts

In 1967-68 the water availability in the town was 63 litres per person per day and the water was supplied six hours a day. The per capita water availability in the city has, however, dropped to less than 42 liters.

According to the Jal Sansthan officials the Kosi and Matela pumps supply 3 MLD of water to the city daily, but the reality is not this. At Matela Jal Sansthan has established three pumps with capacity of 1700 lpm, 800 lpm and 600 lpm. Never do all the pumps operate together and none of the pumps works for more than 6 hours. In the rainy season the rise in the water level causes problems and the Kosi pumps have to be shut off. Frequent power cuts and voltage fluctuations further constrain the quantum of the water supplied.

Despite the fact that the water supply scheme for the city of Almora is causing the loss of more than 80 lakh rupees annually to the Jal Sansthan the department is reluctant to invest upon measures to stop losses through leakage in the supply lines, that have outlived their normal life.

Favourable interplay of geology and precipitation regime provide a large number of natural springs in this region and it is believed that the Almora hill once had 360 *naulas*. This number seems little too exaggerated and is not authenticated by available records. Between 1563 and 1880 the expansion of the township was dictated by the availability of water. Land surveys of Almora were undertaken in 1862, 1904, 1944 and 1966. The survey undertaken in 1862 indicates only 36 water sources (*naula* & *dhara*) in Almora. There is however a noticeable discrepancy in the number of water sources recorded during different land surveys. The 1862 survey recorded 36 sources of water within the city limits and 63 in the adjoining areas. The settlement of 1944-45 recorded 48 sources within the municipal limits of Almora. In the settlement of 1965-66 there were 38 sources in Almora. This settlement was, however, partial and did not cover the entire city. Barren, however, stresses that there were more than 100 springs in Almora in 1842. The detailed studies carried out by Pant (1995) conform to the views of Barren and have recorded 99 springs in the city and its vicinity. Of these 69 sources were observed to be productive while the remaining have either dried up or have been covered by debris. Most of the springs of the town are located along four curved spring lines covering the town with one semicircle each on both the east and west sides of the hill and there is considerable similarity in the locations of the springs on both sides of the hill which indicates possibility of a common recharge areas along these drainage lines.

The springs of Almora town can be broadly classified as conventional open pit type with slate roofing (*naula*) and as closed tanks with piped outlet (*dhara*). Protected from external surface contamination, latter is mainly used for drinking purposes while the former is used for other domestic uses.

Till 1563 all the water requirements of the city were being fulfilled by the natural sources of water. Water scarcity was felt in this mountain township after 1860, when some springs recorded depletion in discharge. The drought of 1884 caused almost all the springs to dry up. Consequent to this the Baldhauti scheme was made in 1884-85. Shail, Shai Devi and Kosi schemes came up later. At present this cultural centre of Kumaon faces severe water crisis and despite several schemes the quantum of water being supplied is far below the civic standards. Rise in population (Fig. 6.1) has no doubt enhanced pressure upon the existing infrastructure, but the lack of vision and managerial skills have worsened the scenario. In the previous few decades the drinking water schemes have induced the people towards neglecting the traditional water sources that once used to satisfy their needs. The state administration also did not pay due attention to these.

The natural springs of Almora have the potential of considerably lightening the ever increasing burden of Jal Nigam/Jal Sansthan. Discharge of Laxmeshwar dhara (Fig. 6.2) that is regularly used by the surrounding households supplements the foregoing assertion.

To get rid of the wastewater drains were first constructed in the city of Almora in 1894 but due to the paucity of water most of these drains remain unclean. During the initial stages of development domestic wastes were collected manually for disposal in open pits. With the development of water supply schemes in the colonial period, concept of septic tanks came into practice for partial treatment and disposal of sewage. So far, an efficient sewage collection and disposal system has not been developed for the town and it influences the surface and groundwater resources of the catchment in terms of quality that directly affects the hydrology of the area.

The domestic wastewater also reaches fresh water sources and pollutes it to a great extent (Goel 1980). In Almora, unplanned growth of the township, without proper collection and treatment of sewage has created real threat of pollution of natural springs located around the town.

There exists no provision of treating the water being supplied to the masses and the quality of the water is dubious. Water quality data of the springs indicate that the concentration of nitrates is beyond the permissible levels of 10 ppm, rendering the water unfit for human consumption (Kumar & Rawat 1996). It suggests leaching of nitrogen from domestic wastewater. The presence of organic matter and coliform bacteria in protected spring waters further indicate groundwater pollution. The concentration of the chlorides in the spring water ranges from 15.48 ppm to 140.52 ppm; higher chloride levels being observed in the springs located in the lower part of the town. The springs located at lower altitudes are thus more polluted with nitrates and organic matters (Kumar & Rawat 1996).

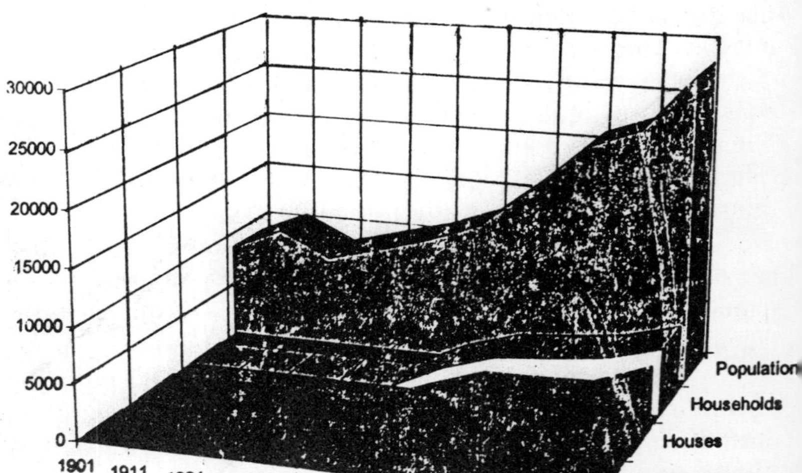

FIG. 6.1: Number of residential houses, households, population and population density of Almora

Sources: Census of India, *District Census handbook*, 1951, 61.
Census of India, *District Census handbook*, XB, 1971, 81.
GISTNIC, Almora, 1991.

In case the surface water gets sufficient treatment within the soil layers, the quality of groundwater would not be degraded. But in the hills, inherent structural weaknesses of the rocks (fractures, cracks and joints) cause shortcircuiting and untreated wastewater often enters the groundwater body to pollute it (Thomas & Backford 1982).

Drinking Water Schemes for Almora over the years

In the time span of 1885 to 1992 four water supply schemes have been implemented for providing solace to the people of Almora (Fig. 6.3):

— Baldhauti scheme
— Shail scheme
— Shai Devi scheme
— Kosi scheme

Baldhauti scheme has ceased to exist due to the drying up of the sources, while Shail scheme has been diverted to rural areas.

Baldhauti scheme

After 1850 Almora faced severe droughts and the drought of 1884 prepared ground for the first water supply scheme for the township. Five sources in Simtola and Baldhauti forest area were subsequently tapped and the water from these was brought to Chinakhan tank from where it was distributed in the city. This scheme fetched 81,720 litre of water daily but the tapped sources started to show deterioration in yield within 25 years of the commissioning of the scheme. In 1899 this scheme was providing 45,763 litre of water daily. After the commissioning of the Kosi scheme in 1951 this scheme died out in the lack of proper maintenance.

Shail Scheme

This scheme was materialised with the contributions from

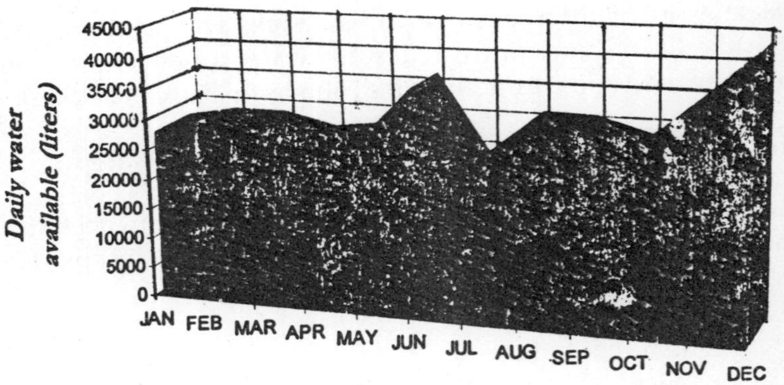

FIG. 6.2: Daily water available from Laxmeshwar dhara in litres (After Pant 1995)

the populace of Almora and the surrounding areas. Tapping water of the three sources in Shail village water was brought to Almora by a 4 km long pipeline. The work on this scheme started in 1884 and ended in 1905, in three phases. This scheme provided 2,35,353 litre of water daily to the city. In 1975 the water from this scheme was diverted to rural areas. At present this scheme caters to the needs of the surrounding villages. The discharge of the three springs has in the meantime deteriorated from 163 lpm to 60 lpm in 1990.

Shai Devi scheme

Seven sources were tapped in Shai Devi and the water was brought to Kankar Kothi tank through a lengthy pipeline. Despite reluctance, severe drought of 1931 forced the municipality to take charge of the scheme in 1932. The discharge at that time was mere 14 gallons per minute.

Kosi scheme

Due to political interference the Almora municipality was

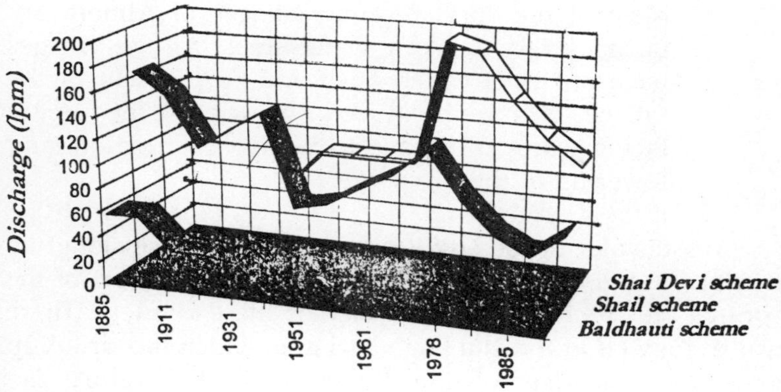

FIG. 6.3: Fluctuating discharge of the three schemes supplying water to Almora township

forced to drop gravity fed Dol scheme that had the potential of delivering 9,80,640 litre of water daily to the town and the Kosi scheme was commissioned in 1975.

Facts

— Water sources are available on both the eastern and the western slopes of the city.
— Most of the sources of water are along the rivulets.
— Most of the sources are becoming seasonal due to the neglect of the recharge area.
— In the densely populated areas the water is polluted, while in the sparsely populated areas the water is fit for human consumption.
— The *naulas* have been neglected for the previous three decades. No efforts have been made by the people or the administration towards their restoration.
— Unplanned drainage in most areas is threatening the purity of the *naula* water.
— In 1985, 791 out of 931 villages of Almora tehsil, 1231 villages of Ranikhet tehsil and 704 villages of

Bageswar tehsil were severely affected by drought.

— Most of the 3,193 revenue villages of Almora are facing severe drinking water crisis. All the villages have not been covered yet and 70 per cent of the villages that were covered by various schemes are facing water crisis due to faulty construction, and lowering of the discharge.

Wastewater of the city is even today let loose along the rivulets, not giving a scant heed to the fact that most of the springs are situated in the vicinity of these rivulets. There was a dugwell in the Narsingh bari area which has dried up presently. Scientists of the Vivekananda Laboratory (an ICAR unit) after testing the water of five drains considered the water to be highly contaminated and unfit even for agricultural purposes. Chemical tests carried out in 1990-91 showed that only a few springs of the city have water fit for human consumption.

Local bodies have no doubt formulated rules for protecting the water of the springs. In 1939 Almora Municipality formulated laws according to which washing of clothes was prohibited except at four places so specified. A fine of Rs.20 was to be imposed on anyone found violating the rules. Interestingly not a single person was fined under the preview of these rules.

Almora Municipality through its notification 931-5-9-B-440-55 dated 23.4.1957 made it mandatory for the Board to get the water quality tested from time to time. At the same time to ensure good quality water to the masses of Almora, prohibition was imposed upon dumping of any polluting material and on the burning of dead bodies upto two kilometres upstream of the Almora water pump at Kosi. Data on water quality could, however, not be accessed and it is dubious if the quality of the water supplied is regularly being monitored.

Naulas were regularly used by the people before 1951. After the Kosi lift scheme most of the households got water connections and the detorioration of the naulas was set in. These naulas if managed and maintained properly have the potential of feeding a large population even today (Fig. 6.4).

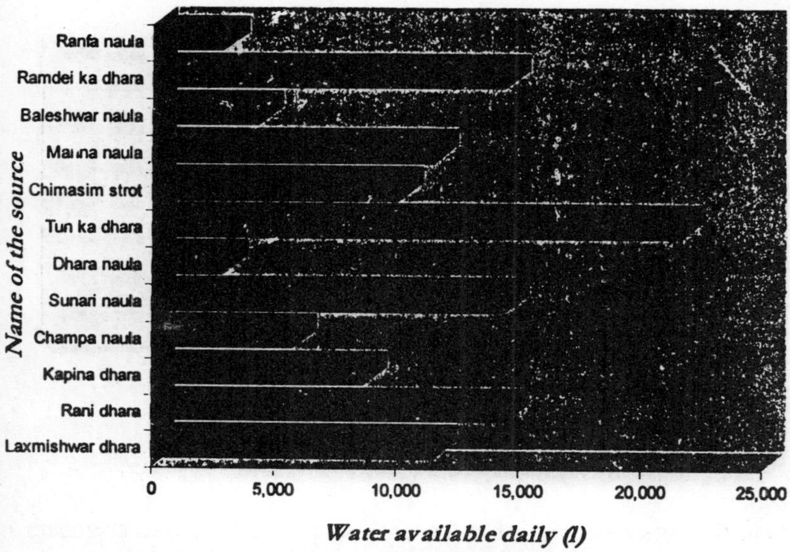

FIG. 6.4: Discharge of some *naulas* of the city of Almora

At present about five thousand people of the city are regularly using the water from natural springs.

Water Supply Scenario in Almora

In the year 1991 the population of Almora was 25,993. At the rate of 150 litre per person per day 38,98,950 litre of water was needed daily. 12,396 beneficiaries to the drinking water scheme of Almora reside outside the municipal limits and there are around 6,000 transitional persons. This escalates the daily water requirement figures to 66,58,350 litre. Industrial and organisational demand is estimated to be of the tune of 10,00,000 litre per day. In the year 1991 Almora township thus needed 76,58,350 litre of water daily. At present the demand is around 9 MLD while the existing water supply schemes have the potential of supplying far less than 5 MLD (Fig. 6.5). Non-availability of electricity and voltage fluctuations render this figure still lower.

The demand of water goes up steeply in summers due to the inflow of tourist traffic, but the supply position

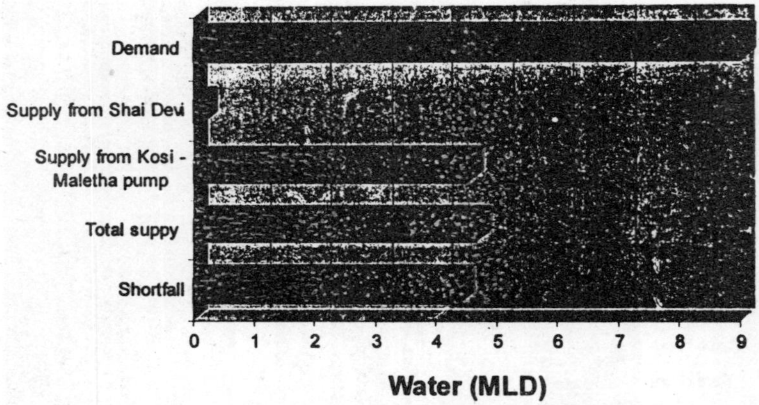

FIG. 6.5: Water supply statistics of Almora township

detoriorates because of the frequent power failures together with the reduced discharge of the streams. The data on the water storage capacity (Fig. 6.6) reflects upon the wisdom of the authorities in keeping the reserve water supply on the lower side.

There is a proposal of lift water scheme from Kosi that would supply Almora with additional 2.5 MLD of water. The wisdom and sustainability of lifting the water from Kosi in itself is a debatable issue. Moreover, the proposed scheme would not even be able to fulfill the present demand for water and by the time it becomes operational the demand would have zoomed so high that it would virtually become useless.

Causes of Water Scarcity

— Reduction in the recharge zone area. Urban development is responsible for landuse changes. The forest and agriculture areas are increasingly being transformed into concrete structures, paved roads and metalled pathways. These impervious structures influence the hydrological cycle by increasing the runoff as percolation of water reduces considerably. Massive drive for replacing

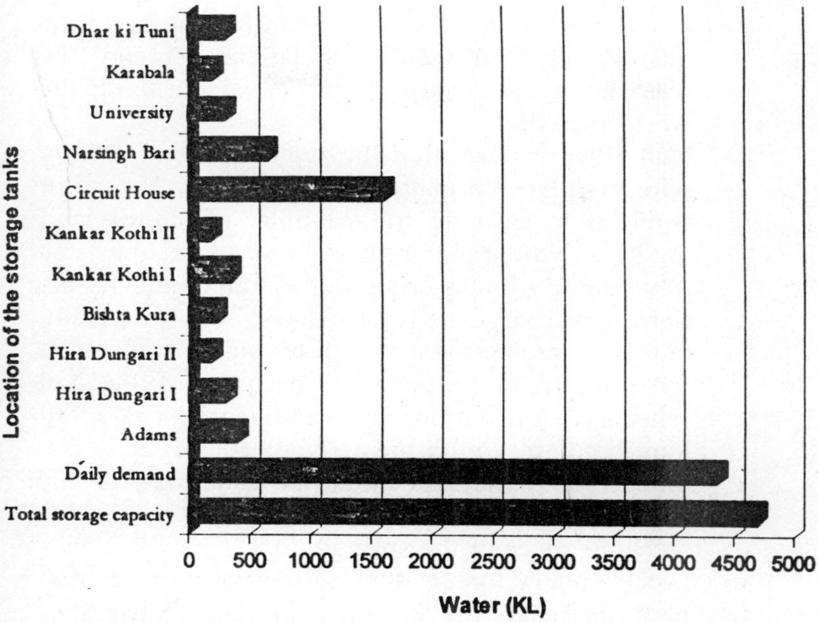

FIG. 6.6: Water storage facilities at Almora together with the total daily demand

the traditional slate paved pathways with concrete in the city needs reconsideration. The concrete paved paths besides influencing the percolation of water also enhances the burden of the municipality upon regular maintenance. Almora is known for extensive use of slate that reflects upon the rich cultural heritage of the city. In the time when the mankind all over is busy preserving its heritage, the decision to replace slate with concrete needs rethinking.

— Construction activity and unplanned dumping of the debris. The natural drainage channels are often blocked by the debris that has the potential of triggering heavy landslides and floods during heavy rains. This type of flooding was experienced in Almora in September 1993. Huge losses were reported by mudflowing and landslides. This aspect of changing runoff pattern can be considered a major problem related to urban hydrology of the hill towns. Stringent regulations should be formulated for ensuring proper disposal of the waste material.

— Pant (1995) asserts that the replacement of oak by pine has played a major role in reducing the water availability. He cites the examples of Simtola and Baldhauti where pine replaced oak and at the same time water sources have shown noticeable reduction in discharge and some have even dried out. Similarly, replacement of oak by pine in Kalamatia forest led to the reduction in discharge of the Shail scheme. Afforestation of the indigenous oak-kafal-burash complex with the active participation of the villagers and students should be undertaken in the barren slopes in the region, particularly in the recharge zones of the springs.

— Lack of policy for the development of pastures that play a crucial role in enhancing the recharge of groundwater. This is stressing the existing forests.

— Lack of any programme to protect the traditional water harvesting structures and to integrate them into the routine preview of Jal Nigam/Jal Sansthan.

— To add to it treated drinking water is being used for irrigation, building construction and other non domestic purposes. A large quantity of water can be saved by reducing the consumption on irrigation and industrial activities.

7

WATER AVAILABILITY IN CHAMBA BLOCK

Situated on the Rishikesh—Gangotri highway, Chamba is a major economic centre of Tehri Garhwal district that enjoys the dubious distinction of being severely water-starved township. Almost all the water supplied in the township is pumped from the nearby Henval river and is far below the required quantum by all civic standards.

Chamba block falls in the sub-tropical climatic zone of Himalaya, with average annual rainfall figures of around 1200 mm. The details of the meteorological data are given in Table 7.1 and based on these some graphical representations are presented below (Fig. 7.1).

The maximum temperature recorded in the previous decade is 26.7°C in June while the lowest recorded temperature is 2.5°C in January. During winters the higher reaches experience solid precipitation.

The relative humidity is more than 45 per cent all through the year and 61 per cent of the annual precipitation is confined to the four months of June to September, while the minimum rains are experienced during October—November.

Water availability in four selected villages of Chamba block was studied in detail. All these villages; Lawadhar, Than, Paturi and Bidon fall in the catchment of Henval river and are far off from the road head. None of these villages has access to any of the perennial streams in the vicinity. Incidentally, all these villages face the common problem of water scarcity, but the problem is more acute in Lawadhar and Paturi that do not have natural springs in the vicinity.

TABLE 7.1: **Meteorological parameters recorded at the Hill Campus Ranichauri. The data presented is the average of ten years (1985-95)**

Month	*Temperature (°C)*		*Relative Humidity*	*Rainfall*	*Sunshine*
	Maximum	*Minimum*	*(%)*	*(mm)*	*(hrs/day)*
January	11.9	2.5	55	61.8	5.3
February	13.4	3.0	59	84.4	6.0
March	18.1	6.4	52	91.3	7.0
April	22.9	10.4	42	59.3	8.8
May	25.6	13.8	45	84.9	9.7
June	26.7	16.5	57	113.6	7.7
July	24.7	17.0	82	273.5	4.6
August	24.0	17.1	85	251.1	3.9
September	23.3	15.0	83	123.7	5.7
October	21.1	10.3	63	27.5	8.6
November	17.1	6.1	55	14.7	7.0
December	13.3	3.3	59	55.5	5.5

Facts

— Water is supplied to Chamba and the surrounding villages through lift water scheme from Henval and Jijli gad.

— Total of 1.8 MLD of water is supplied by this four stage lift scheme.

— The four pumping stations at Nagni, Chopriali, Sabli and Pantnagar have storage capacity of one lakh litre and are powered by three motors of 100 HP each.

— Local distribution tanks of varying capacities are located at Ranichauri, Thang, Sabli, Dargi, Badon, Saunkoti, Badshahi Thaul and Chamba.

— To add to the water demand of this water depleted

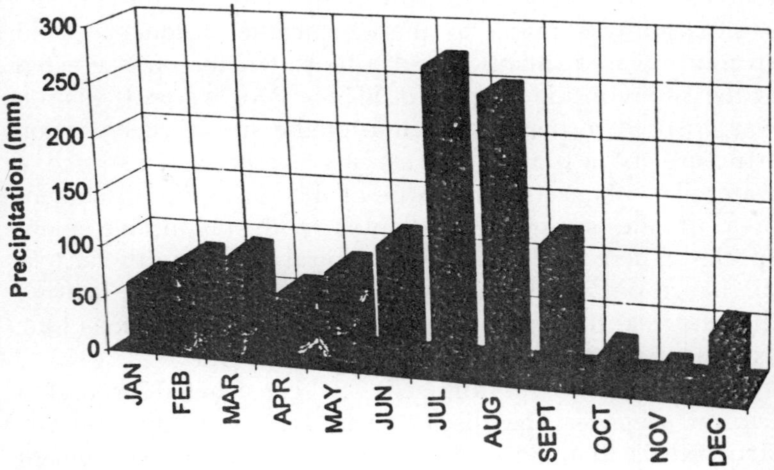

FIG. 7.1: Monthly rainfall variation at Ranichauri in Chamba block

zone new township of New Tehri has come up in the surrounding.

Almost all the agricultural land of these villages is unirrigated. Except for the upper reaches of Lawadhar Paturi where mixed *oak-burash-kafal* forest is still thriving most forest land has a monoculture of pine and the crown density is low. Wood is the main source of energy in these villages and the ever increasing demand of the forest produce has degraded the forest close to the habitations.

Lawadhar does not have any natural seepage of water in its vicinity. To the west of Raturi Khala above the fields there existed a perennial spring. A tank was constructed at this site to harvest the water. Prolonged neglect of the spring due to the breakdown of the social mechanism for maintenance led to reduced discharge and the 1991 earthquake sounded the death knell. Only moisture was initially observed at the site. Water at this source is observed to ooze from the shear zone and dismantling of the old structure together with trenching revealed appreciable discharge. Trees for Life/HIMCON (Ranichauri) workers found it to be a potential site for demonstrating spring rejuvenation strategy. With the cooperation of the villagers a strategy was made to implement the vision. The old structure was dismantled and a new structure was erected at the site with the support of IGSSS. People's participation was ensured in the construction of the spring rejuvenation structure and it is at present serving as a perennial source of water. To enhance the recharge of the spring the catchment area of the spring was planted with favourable plant species. There also exists one natural spring with meagre discharge outside the village boundaries (Plate 7). During the summers the females of the village have to walk as long as 5 kilometres to fetch water from Chani gad.

In Paturi a *koo* (dugwell) is constructed around a natural seepage. The discharge is, however, low and the structure is ill maintained. There exists a defunct cement lined water harvesting structure at the slope break in Paturi. The water harvested in this structure used to augment irrigation. Three *khals* are observed in the forest boundary above the Paturi village at the slope break. Grazing animals use the water collected in these traditional structures (*khals*) and these also augment recharge of ground water. TFL/HIMCON has built a few more khals in the area and a series of water harvesting structures at favourable sites in the seasonal stream. In times to come these are envisaged to augment water availability in the region. Entire village population is satisfied and is cooperating in these efforts in which their ideas and suggestions are honoured.

There are two natural springs within Bidon. During the field study it was gathered that there existed one more spring at the slope break above the flat lands. This has now dried up and the levels of the existing two, together with their discharge, have gone down consequent to various tectonic activities. At the site of the spring the mechanism for separating used water from that for animal use has been disrupted due to the lowering of water level (Plate 8). There existed a gool that was fed by the overflowing spring water; lack of maintenance has, however, led to its deterioration. Than also has two natural seepages in its vicinity. These are

PLATE 7: Dhara near Lawadhar with meagre discharge

LATE 8: Watertable at Bidon has been lowered during the 1991 Uttarkashi earthquake and none has cared to renovate the old architecture so as to provide separate space for washing and for animals

situated along the slope break above the village. The efforts made in the past towards enhancing the water availability of these springs have taken their toll and these springs have turned seasonal. At both the locations concrete tanks have been erected and the water channel has also been lined with cement. It seems that the passage of water channel has been disturbed by the efforts made in the past.

At present the water related requirements of the people are met by the tap water that is supplied by lengthy pipeline. The individual households of the villages do not have personal tap connections and the entire village has to be fed from common connection (1 in Lawadhar, 7 in Than, 4 in Bidon and 2 in Paturi) connected to the tank. This creates great inconvenience to the villagers. The condition of the tanks made for storing the tap water is also deplorable and needs immediate maintenance.

Due to various technical and other reasons the tap water supply often fails, specially during summers. Under such circumstances water has to be brought from the springs situated at far off places. This obviously enhances the routine workload of the females.

The water requirements of these villages are estimated (Table 7.2, Fig. 7.2) on the basis of the demography. This data shows that drinking water requirements of these villages are comparatively low and bulk of the water is needed for other household purposes and for feeding the animals.

Rainwater harvesting from the roofs has enormous potential in this area. The roofs are generally slanting and traditionally made of slate; water can be easily collected from these roofs making use of metal sheet drains. This water can be collected in LDPE tanks or ferrocement tanks erected on the surface. The rainwater harvested thus from the rooftops shall be sufficient to meet most of the daily household requirements. Moreover, this water would be available at a short distance. Lawadhar, Paturi, Bidon and Than have 30, 50, 45 and 71 houses respectively. On an average the covered area of each house is of the order of 40 sq m. Estimates of the water thus collected from each roof during different months are given in Table 7.3. This data

TABLE 7.2: **Daily water requirements of the four studied villages of Chamba block**

	Population	*Cattle population*	*Daily water requirements*			
			Drinking @ 7 litre	*Other household purposes @ 30 litre*	*For cattle @ 20 litre*	*Total (in litre)*
Than	415	242	2909	12450	4840	20199
Lawadhar	190	46	1330	5700	920	7950
Bidon	338	87	2366	10140	1740	14246
Paturi	338	154	2366	10140	3080	15586

TABLE 7.3: **Estimates of the water likely to be harvested from one roof during different months of the year**

Month	*Monthly water harvested from one single roof (in litre)*
January	2472
February	3376
March	3652
April	2372
May	3396
June	4544
July	10940
August	10044
September	4948
October	1100
November	588
December	2220
TOTAL	49622

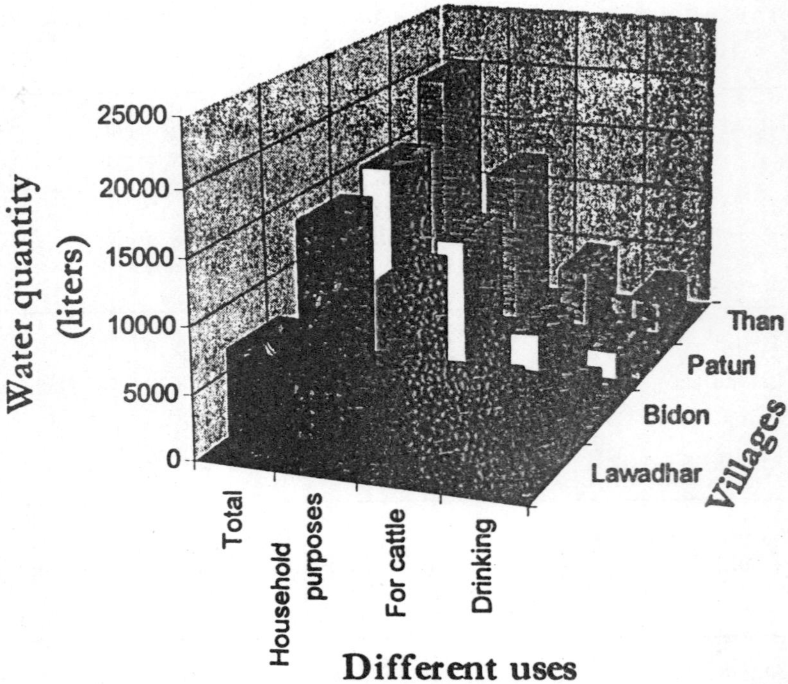

FIG. 7.2: Diagrammatic representation of the daily water requirements of the four studied villages of Chamba block

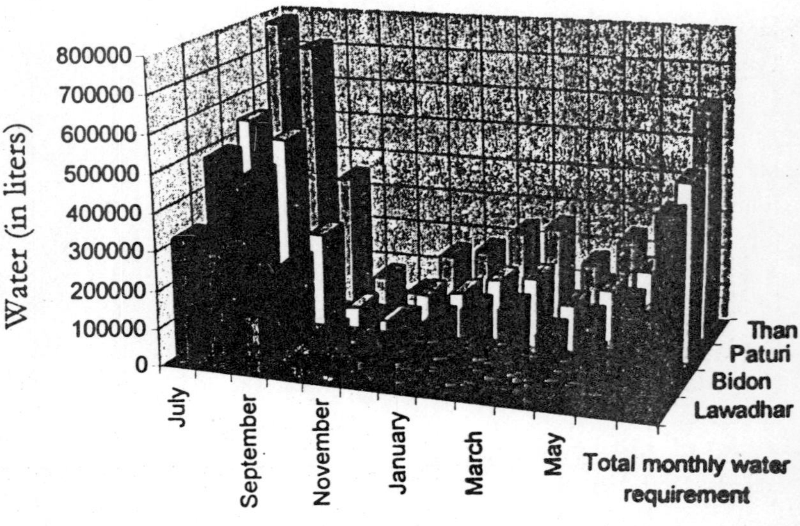

FIG. 7.3: Water that can be harvested from the rooftop in the four villages studied

shows that except for the months of October and November the water harvested would be fulfilling majority of the house-hold requirements (75%) for the family of 5 and having two cattle. If the household does not have cattle its roof is potent enough to fulfill its water related requirements. Based on the actual demographic figures the water that can be harvested from the rooftop in different villages is presented in Table 7.3.

TFL/HIMCON is doing every effort towards conserving the available water in the villages of Lawadhar and Paturi. It has built a series of rainwater harvesting structures (ferrocement and LDPE tanks) in these villages (Plate 9).

8

THE DOON VALLEY

Inadequate recharge resulting in lesser discharge of the springs and wells and increased demand of water is creating scarcity of water in the township of Dehradun. The groundwater play an important role in solving this periodic water crisis and it also serves as a source of uncontaminated water. More than 50 tubewells and 250 dugwells are in existence in the valley. In the decade between 1972 and 1982 groundwater level in the Dehradun city has dropped by more about 30 metres (Thadhani 1985) and overexploitation of groundwater may lower the watertable further and reduce the saturated thickness of the aquifer.

Laduse changes (Fig. 8.1) are considered to have contributed largely to the reduced recharge and consequently reduced discharge of the natural springs.

Facts

- The population of Dehradun district rose from 1,44,000 to 7,58,000 in the hundred years from 1880 to 1980.
- In 1980 the population density had reached 246 people per sq km which is very high a figure for the mountain valley to sustain.
- The available arable land in the district has shrunk from three tenth of a hectare per person in 1880 to one tenth by 1980.

The Lesser Himalayan rocks exposed to the north of

Dehradun, though sedimentary, have a very low inter-granular porosity and are characterised by fissures, fractures and joints. The groundwater in this zone occurs largely as disconnected local bodies in favourably perched aquifers under both confined and unconfined conditions and also in zones of jointing, fracturing and faulting. Relatively flat areas and gently sloping grounds characterised by deep weathering, such as hill tops, ridges, saddles, spurs and bulges of old landslide debris, river terraces and fluvial fans form the recharge area while steeper hill slopes, 1st or 2nd order streams at slope breaks and scarps of fans are observed to be the sites of discharge. The upper portions of the catchment areas are saucer shaped amphitheathers giving rise to large underground bodies of perched water. The springs and seepage issue out from all sides of the individual aquifers at different elevations. The springs in the rocks of secondary porosity show great variability in yield even within short distances.

The Krol Formation (limestone terrain) exposed to the north of Dun valley is characterised by cavities and solution channels oriented along WNW-ESE and NW-SE. Selective

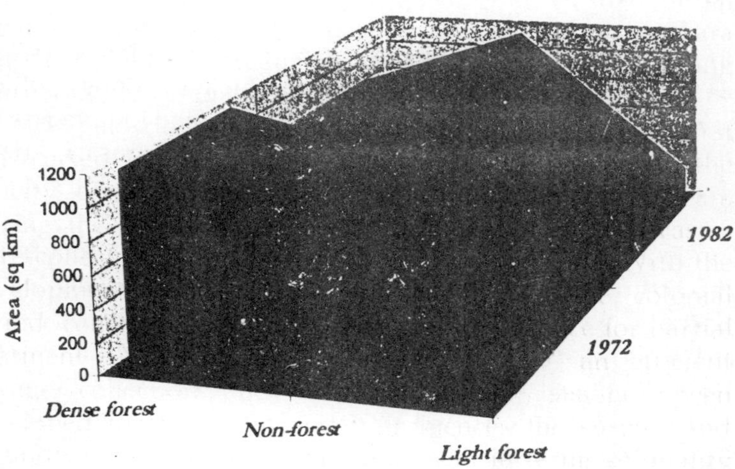

FIG. 8.1: Landuse/Landcover changes in the Doon valley from LANDSAT MSS data (After Tiwari & Garg 1995)

weathering along fractures and joints has created network of underground water courses leading to almost complete lack of 1st and 2nd order perennial streams on the surface. The yield of the springs in the karst belt is very high; of the order of 1,73,000 to 5,36,000 l/day.

The Doon valley is endowed with very productive aquifers within the Doon Gravels that constitute alluvial fill in the valley. In Doon valley number of water table springs are present along the valley floor and along the terrace and fan scarps contributing to the base flow during the lean periods. The discharge of such springs is usually high and vary from 2,000 to 25,07,000 l/day. Available lithologs of the bore holes and the exposed vertical sections indicate that the horizons comprising boulders and gravels set in coarse sandy matrix are the main water bearing horizons. These horizons are encountered more than once and are separated by clay horizons having few small pebbles and gravels that act as aquiclude (Bartarya 1995). The water table is shallowest in the central part of the valley, in the subcatchment of Asan river and deepest close to water divide of Asan and Song rivers near Kaulagarh.

Bartarya (1995) has described various types of springs encountered in Doon valley. These are:

Fault-lineament related springs : Fault planes (weak planes in the rocks) act as pathways along which groundwater moves and rises upto the surface to give rise to these springs such as at Kaphlani, Chhamsari, near Hathipaon, and Talai.

Fracture-joint related spring : These springs are formed in fractured and jointed hard sedimentary rocks such as in Masrana, Anand Chauk-Kathu-ki-chail, Dibli, Dwara, Gwar and Buraskhand area. These springs originate either along the hill slopes or along stream beds wherever the local water table is intersected by the fractures and joints or by the ground surface.

Colluvial springs : The upper slopes of the recharge areas of the springs are generally low (8-12°) which provide good hydraulic gradient. Such springs occur at Selagaon, Suakholi, and Ghaisain.

Fluvial deposit related spring: Because of difference in porosity and permeability of the deposits and underlying bedrock, the groundwater is discharged as springs through the contact of alluvial fan and terrace and also at the contact of fluvial deposits and the bedrock. Examples are many in the Tal valley, upper reaches of Song, Surna, Nun, Baldi and Bindal.

Karst springs : The water emanating through cavities and underground channel in calcareous rocks (e.g. limestones) such as in Sahastradhara and Durmala area. The cavities and channels are mostly joint and/or fault controlled. The discharge is higher than the other types.

Post Independence Efforts?

— Nala Pani springs were first tapped for Dehradun in 1889.

— Kolukhet was tapped for Mussoorie in 1895.

— The Murray springs were tapped for Mussoorie in 1908.

— The Mackinnon springs were tapped for Mussoorie in 1913.

— Bhilaru was tapped for Mussoorie in 1925.

— Koltikhala pumps were commissioned for the Landour area in 1929.

— Nothing significant seems to have been done since then. Part of the problem lies in the fact that in 1971-72 the responsibility for water supply ceased to lie with the Municipal City Board. It was passed to the Garhwal Jal Sansthan that was responsible for supply of water and the Jal Nigam that was responsible for installation of new schemes. To add to it rampant corruption and mismanagement, lack of responsibility and political will keep the solution elusive.

Facts

Population of Dehradun is around 2,90,000 (approximately 58,000 households). Bisides, the city also harbours three groups of floaters as well; seasonal immigrant labourers settling in the slums, tourists and students. This escalates the city's population to more than 3,22,000 in dry season.

The city of Dehradun is spread over 32 square kilometres and its sources of water are 30 odd tubewells, with a maximum capacity of 52.12 MLD, and the Bindal river, providing upto 9 MLD. A total supply of 61.12 MLD should, therefore, be expected during regular season. This is equivalent to 163 litre per capita per day for the total population of 3,75,000 (2,90,000 in Dehradun and 85,000 in the adjoining areas).

Based on their pumping rate of 59,550 lpm, the total regular season capacity of the city's 30 tubewells should be 64 MLD. One of the tubewells is however abandoned since 1986 and most suffer from deficiencies in pressure gauges, depth gauges or booster pumping stations. Electric breakdowns during the dry season further affect the output.

On an average the tubewells of Dehradun are 120 m deep with the depth of groundwater around 70 m during the regular season and 110 m during the dry season. Four of the tubewells are however less than 60 m deep, and are therefore, inoperable during the dry season. This alone constitutes 7 per cent loss of daily production during the dry season.

Four water treatment facilities are operational in Dehradun, but their capacity totals only 40 MLD, well short of the daily water production. Only six of the tubewells have working chlorinators. A large population is, therefore served untreated water.

Water is distributed from the city's 11 overhead tanks for six hours a day during the regular season.

The sewer line serves a quarter of the city's population. Most of the residents (66%) discharge their wastewater into open. The existing five sewage pumping stations supposedly pump sewage to a nearby river; however, three out of five pumping stations do not function properly. To add to it all, the city does not have any wastewater treatment facility. The lack of proper sewage lines cause groundwater contamination and sewage infiltration into low pressure water supply lines.

Following rains, water starts percolating down through the weathered material and saturate pore spaces in soils and rocks. The recharge of groundwater tends to increase the amount of groundwater storage and raises the pressure in primary and secondary pores. This result in increase in the discharge of water through springs, and thus its quick outflow. High rates of discharge lower the water table and its gradient, and reduce pressure in pore spaces. This alteration of recharge and discharge is the cause of seasonal, local and short term fluctuations.

Water Scarcity in Mussoorie: The Queen of the Hills

The town of Mussoorie needs to pump 30 lakh gallons per day, or 1.5 lakh gallons per hour for 20 hours a day to meet the requirement of 1,00,000 people in the peak tourist season at the rate of 30 gallons of water per person per day. As against this the town's pumping capacity is barely 75,900 gallons per hour and owing to poor maintenance of the pumping stations, it does not pump more than 54,000 gallons per hour for a maximum of 16 hours a day. The town needs to tap known new spring water sources from which it can make up the current shortfall of 1,00,000 gallons of water per hour in a reasonable time frame and at a reasonable cost.

Back up water storage needed for two days water supply is 60 lakh gallons. Out of Mussoorie's installed reservoir capacity of 45 lakh gallons, 23 lakh gallons are likely to go out of use at any time because of poor maintenance. The estimated requirement of water for Mussoorie town does not take into consideration the needs of the lower, surrounding villages. The town runs on outdated pumps and the open collection chambers in streams, from where water is pumped to the town, are full of debris, leaves and stones, which choak the pipe outlets of the pumping stations. The approach to pumping station is difficult to negotiate and no one goes there if they can help it. No one seems interested in setting right these bottlenecks, the cost of which is less than 1 crore, to double the water supply of the town. The interest is in a new and grandiose schemes costing several crores of rupees, which have little hope of early sanction or implementation, because of the expanses involved. One also feels that small repairs are deliberately ignored to create a demand for more expansive schemes.

LOST GLORY OF GANGAR

Gangar is situated in the Bal Ganga valley in Tehri Garhwal district. The village was once habitated by happy go lucky persons who owned vast tracts of fertile land and large animal stock. The animals thus provided manure to both the farm land and the nearby forests where these grazed. This maintained the fertility of the lands and the people were gifted with the bounty of biomass production. The quantum of the surplus production can be assessed from a number of magnificent *kothars* (granaries) that the village has. As many as half a dozen were observed in the village. These are small structures housing the chambers to store the agricultural produce. Each *kothar* has as many as nine separate compartments to store different types of grains. These are made up of wood. The storage capacity of each compartment is around one cubic metre. The houses are double storied with slanting roofs and wooden carvings on the doors and the windows. Huge gate was observed in one house together with the space for the watchman to stand.

Though there are three *dharas* on the lower reaches of the village natural springs are not present in the vicinity of the village. This handicap was, however, beautifully overcome by properly channelising the water of Gangar Gad. A checkdam was constructed across Gangar Gad at appropriate elevation and *gools,* the remanent of which are still visible, were made to take water to far off fields and households. The labour of the people fetched rich devidends of agricultural produce and prosperity of the masses enhanced.

The prosperity of the village led to the ancestors

acquiring land interests along the banks of Bal Ganga at Baheri. These lands were maintained by the people of the village who shifted to this place during sowing and harvesting periods and accordingly temporary residence were made at this place.

With the roads came materialism into the region and the people were more and more inclined towards comforts. This led the people to shift their residences to Baheri permanently. The lands of Gangar were still maintained by seasonal migration while the fodder and fuel-wood was supplied from the jungles surrounding Gangar. The menfolk thereby added to their comforts while the workload of the womenfolk was enhanced by manifold as they had to devote much more time and energy for collecting fuel and fodder, the distance of which increased with migration. The lands were also deprived of regular supply of the manure. This was reflected in the deteriorating fertility and productivity.

With the passage of time the process of migration was complete and the seasonal migration to Gangar was discontinued by most households. Some, however, continue to sow the lands seasonally without caring much for the returns while most of the lands were left barren. It was only the interests in the surrounding jungles that the females visit this place regularly. The menfolk visit Gangar for ritualistic purposes at the shrine of *Narsinha Devta*. At present Gangar is habitated by a few low caste families that were from the very beginning pushed towards marginal lands and, therefore, could not muster enough surplus to purchase interests in land outside. Moreover, their vulnerable social position prevented their migration.

With gradual increase in population the land at Baheri, though fertile, is finding it hard to meet the increasing food grain demand. Fragmentation of land has further deteriorated the scenario. Moreover, the drudgery of the women has led to increasing resentment, though subdued. The forests around Gangar are also under pressure due to the surmounting fodder and fuel-wool demand. When contacted most households want to shift back amongst the woods, to Gangar. But the state of lands and structures at

Gangar cannot support the influx of people. Most of the houses need immediate repair. The nutrient cycle of the land has been disrupted and thorough work needs to be done to restore its productivity. The old system of irrigation has been disrupted and even the drinking water taps are non-existent.

Within the Gangar Gad watershed there are no reported seepages of water but the water entering the surface would certainly be adding to the stream flow besides enhancing the soil moisture. Prolonged neglect and exploitation of the recharge zone has led to the decreased recharge. This has adversely affected the soil moisture in the area. In order to augment the soil moisture trenching together with plantation is required.

Old arrangements that once used to provide water for irrigation and household purposes to the population of Gangar have turned non-functional over the years. State of water availability in Gangar is owed to the prolonged neglect of the traditional systems.

10

DRINKING WATER AROUND NAGTHAT

Jaunsar Tribal belt has not been studied in great detail by environmentalists, but has always been the centre of attraction amongst the anthropologists. This region, because of polyandry, is much romanticised than is actually understood. Situated in the heart of the Jaunsar, Nagthat (Dabred watershed) in Dehradun district lies between 77°56′ and 78°01′ E longitudes and 30°34′ and 30°37′ N latitudes with average altitude ranging between 800 and 2200 m above mean sea level (MSL) and can be approached by motor road diverging from the Yamunotri highway. Dabred stream that drains into river Yamuna near maror (on Yamunotri highway) forms the principal stream of this region. Small streams/nalas originating in the western part around Singhori, Barat khal and Chhicherar are the major contributors to this drainage system.

As everywhere else in Uttarakhand lot of the women in this region is also miserable. They shoulder the burden of entire household chores besides the work outside the four walls of the home (Plate 10). Though the general health of the children in this region was observed to be good, the health of the grown ups is a cause of concern. This may be due to the addiction to the locally brewed liquor and tobacco amongst both sexes. Once a tool of adaptation, that helped the people to relax after the days hard work, liquor has taken form of a major social evil and is a cause of concern. Besides this the water quality data of the region warrant detailed studies.

Water ranks high amongst the priorities of any human settlement and at the same time dictates the living conditions of the people. In this remote region natural springs are the only cost effective and viable means of fulfilling the needs of these people. At present females have to devote major portion of their time for the collection of water, specially in the summer seasons when the pipelines fail and the supply is severely reduced. It was noted that females have to go as far as 2 kilometres covering the rugged and steep topography to fetch water under such circumstances. Improved availability of water would be providing females with ample time to devote towards other productive activities. This would be improving the social status of the females. Moreover, enhanced availability of water can be utilised for generating energy for diverse uses. Landuse pattern of the region was delineated with the help of the available Survey of India toposheet (Table 10.1).

TABLE 10.1: **Landuse pattern of Nagthat**

	Area		
Type of landArea	*(in Sq. km)*	*(in hectares)*	*Percentage*
Cultivated	13.7325	1373.25	59.4
Forest cover	1.7525	175.25	7.6
Barren land	7.6275	762.75	33
Total	23.1125	2311.25	

Water Harvesting Traditions

Habitations in the hilly terrain have interwoven with them a tale of water harvesting, and this region is no exception. The water harvesting traditions of this region are magnificent and unparalleled, though not studied in detail. At the western edge of the watershed, upon a hilltop, lay the ruins of some historical monument from which the local people proudly trace their descent. These the locals attribute to *the Mahabharatian* king Virat, in whose court the

PLATE 9: Ferrocement tan[k] at Lawadhar (constructed [b]y TFL/HIMCON Ranichaur) collects roof runoff. The beneficiary has ready access to water at her doorsteps

PLATE 10: List of females burdens in hills is endless and there are many a burdens to be shouldered by her besides fetching water

Pandavas took shelter during their *agyatvaas*. Not much of interest could be found amongst these ruins except the remains of a brick lined well.

The nature of the construction together with the provisions created in the hill for defence, certainly indicate that it was constructed by some king, but how authentic is it to associate the same with Virat is a question of debate. The bricks are *Lahgori* which were not constructed in post-Mughal times. It certainly puts the structure contemporary to the Mughals; nothing more can really be said with the available evidence.

It is interesting to note that all the other structures at the hill top are made of stone while the well is lined with the fire baked bricks. In the construction of the bricks very coarse sand has been used and the proportion of clay is significantly low. From its architecture the structure resembles the ones constructed for exploiting groundwater, rather than a simple rain water harvesting structure that is normally to be expected here. It was probably exploiting some perched horizon at the hill top that is made up of phyllitic rocks. The technology employed for identifying the water bearing horizon must have been highly evolved in the region. No traces of this technique could however, be recollected. At present the well is defunct and has been filled up with stones.

Besides this magnificent but defunct structure, there exist a number of natural springs in the region. These seepages are regularly used by the locals for diverse uses. In the vicinity of the seepage provision for drinking of cattle is mostly provided. Most of these springs are situated in the colluvium and the state of the forest in the recharge zone is generally good.

Traditions of nature-man-spirit complex interwoven around the natural springs were observed in the region. A traditional water harvesting structure with a platform in the middle was observed in Rampur village. It is locally referred to as *pokhar* or *pap-ka-dhara*. The platform is associated with *Pap devata* and the water is regarded sacred. Interestingly enough the state government has erected a public stand post at the site of the traditional water source at Rampur (Plate 11). The logic behind this could not be traced and the people in other parts of the village do not have access to other sources of water. The *dhara* near Bisoi is similarly referred to as *devta-ka-dhara* and is associated with local deity Lord *Mahasu* (incarnation of Lord Shiva) whose wooden motif was observed at the site of the *dhara* (Plate 12).

Water source near Lacha (Lacha-4) is considered to be sweet and healthy by the locals. They hold that the temperature of the *dhara* varies inversely with the atmospheric temperature. Water of Lacha-1 and *Gadiakhara* in Mundhan is belived to create stomach disorders.

More than twenty-five springs were located, sampled and studied in this region as against only four depicted in the SOI maps. Most of these springs were noticed on the western side of the watershed. Many of these are however, reported to be of seasonal nature and dry up in the summers. The discharge of the springs was recorded in the field (Table 10.2).

All the villages have tap water supply that often fails and the villagers have to take recourse to the traditional springs. The water for tap water supply is either collected from the springs or from the perennial streams. The condition of the water collection tanks was pitiable with a lot of decaying organic matter. When questioned people in the Duna village confirmed that they do not use it for drinking purposes for which the water is collected from the springs.

This region is gifted with abundant rains but the major portion of it goes waste in the form of surface run off, taking along with it our fertile soil.

As per the data collected in the field as much as 1.2 MLD of water can be made available from the springs for diverse usage, to the population of the area. Besides this, there exists stream water in the area that can well be diverted for agriculture. The irony is that sufficient drinking and agriculture water is not available to the masses, and the spring water is flowing waste in almost all the areas. The productive use of the spring water is as low as 5 per cent, that too from the ones that are situated at

PLATE 11: Development dilemma: Traditional water source has been chosen as the site for community water connection at Rampur

PLATE 12: Religio-magical rites are often interwoven around traditional water sources in the hills. Wooden motif of Mahasu at a *dhara* near Bisoi

convenient locations within the village. The water of the springs can be adequately utilised for diverse purposes by erecting storage tanks at suitable locations and then diverting the stored water to the convenient position through the network of *kuhls*. People's participation in the planning and construction activities should be ensured to make this strategy acceptable to the masses.

TABLE 10.2: **Discharge of the springs around Nagthat**

Water source in brackets is the name of the village	*Discharge*	*Daily water available (litre)*
Upper *Naigar* (Gadaul)	7.3	10,512
Lower *Naigar* (Gadaul)	1.3	1,872
Teday Khadd (Gadaul)	590	849,600
Lacha-0	6	8,720
Lacha-1	43	62,525
Lacha-2	2.5	3,490
Lacha-3	2	2,280
Lacha-4	27	39,198
Lacha-5	3	4,320
Lacha-6	3	4,320
Baikhuber (Gadaul)	1.8	2,592
Hajaka (Gadaul)	29	42,027
Near (Rampur)	2	2,280
Pokhar/Pap-ka-dhara (Rampur)	31	45,024
Upper *Jogranda* (Nagthat)	180	259,200
Lower *Jogranda* (Nagthat)	60	86,400
Duena	7	9,630
Devta-ka-dhara (Bisoi)	54	77,433
Pani-ka-dhara (Sheela)	10	14,400
Periyandu (Kuwasa)	30	43,200
Ukhniyan (Khuna Amal)	13.5	19,440
Prantkey (Khuna)	7.6	10,944
Bai-ka-pani (Kandoi)	3	4,320
Total	1114	1,603,727

The recharge areas of most of the streams are adequately forested, but there is ample scope for constructing *chaals* that have the potential of enhancing the recharge of the aquifers. The water collected in the *chaals* would be available for the grazing animal, and the effect of the same would be reflected in the lower erosional rates.

Physio-chemical Studies

In the present investigations 27 springs and 5 sites for surface water were selected for detailed monitoring through sampling and testing for water quality and data analysis. The main objective of the study was to locate both ground water source (spring) and the surface water (stream/ *khal*/ waterfall) for the continuous assessment from potability point of view, as these water sources are the only drinking water sources for the inhabitants of the watershed. The water samples were collected for six consequitive days between 0800 and 1200 hours and were brought to the laboratory in pyrex glass specimen bottles. Water quality parameters like pH, dissolved oxygen (DO), free carbon dioxide (FCO_2), carbonate (CO_3^-) and bicarbonate (HCO_3^{--}) were estimated on the spot immediately after collection, whereas water analysis relating to the other physio-chemical parameters were done in the laboratory. The temperature was measured with mercury filled Celcius thermometer while the pH was recorded with the portable pH meter. Other parameters like DO, FCO_2, TDS, TSS, total hardness, chloride, Ca^{++} and Mg^{++} were accessed adopting the standard techniques (Welch 1984, APHA 1989). Correction coefficient among the physio-chemical characteristics of both ground water and the surface water sources from different sources were also statistically correlated.

Variation in physio-chemical characteristics of both surface and ground water from different locations are summarised in Tables 10.3 and 10.4. In general it was observed that both springs and the open water located at lower altitudes have different chemical characteristics in comparision to those located on the ridges. The water

TABLE 10.3: Physio-chemical characteristics of the spring water

Water source in brackets is the name of the village	*Ambient temp. (°C)*	*Water temp. (°C)*	*pH*	*TDS*	*TSS*	*TS*	*DO*	FCO_2	HCO_3^-	CO_3^-	*Chloride*	*TH*	Ca^{++}	Mg^{++}
1	2	3	4	5	6	7	8	9	10	11	12	13	14	15
Upper *Naigar* (Gadaul)	11.2	9.4	6.3	225	90	315	7.8	1.5	25	—	21	42	10.4	3.9
Lower *Naigar* (Gadaul)	13.7	9.0	6.4	100	105	205	8.2	5.0	20	—	18	32	8.0	2.7
Teday Khadd (Gadaul)	13.8	12.4	6.4	150	135	185	7.0	2.5	35	—	22	40	11.2	2.9
Lacha-1	12.4	14.6	6.4	135	105	240	8.0	4.0	35	—	24	26	7.2	0.51
Lacha-2	9.8	10.6	6.4	150	110	260	7.4	5.0	30	—	23	24	6.4	1.94
Lacha-3	18.9	13.5	6.4	144	85	229	7.8	1.0	25	—	18	28	9.6	0.96
Lacha-4	18.9	14.2	6.5	185	300	485	7.0	3.0	30	—	30	32	10.4	1.45
Lacha-5	15.5	13.0	6.5	150	105	255	9.0	6.0	20	—	19	24	7.2	1.45
Baikhuber (Gadaul)	19.6	14.2	6.4	425	225	640	8.4	6.0	30	—	27	18	4.8	1.45
Hajaka (Gadaul)	19.5	13.6	6.4	400	90	490	10.4	7.0	30	—	23	28	8.8	1.45

Roagad Khadd (Rampur)	13.5	14.8	6.6	180	90	270	7.8	2.0	40	—	15	46	14.4	2.41
Pokhar/Pap-ka-dhara (Rampur)	17.5	13.0	6.5	85	95	180	7.8	2.5	40	—	24	44	14.4	1.92
Gadiakhara (Mundhan)	19.0	15.5	6.5	225	110	335	9.0	4.0	65	—	18	66	22.4	2.41
Chichrar	19.5	14.0	6.7	200	95	295	8.0	1.5	55	—	16	68	22.0	3.24
Upper *Jogranda* (Nagthat)	22.3	17.6	6.5	125	150	275	8.6	4.0	75	—	33	70	24.1	2.19
Middle *Jogranda* (Nagthat)	22.3	16.8	6.7	325	350	675	7.4	6.0	65	—	32	68	23.2	2.41
Lower *Jogranda* (Nagthat)	22.8	16.9	6.5	225	225	450	7.8	6.0	60	—	33	52	16.8	2.41
Duena	10.8	13.5	6.4	130	175	305	7.4	4.0	25	—	21	24	7.2	1.45
Devta-ka-dhara (Bisoi)	14.8	15.4	6.3	85	300	385	8.2	3.0	30	—	18	24	7.2	1.45
Pani-ka-dhara (Sheela)	16.5	13.5	6.4	185	105	290	6.8	2.0	20	—	25	26	8.0	1.45
Periyandu (Kuwasa)	19.5	14.2	6.7	145	85	130	7.4	3.0	40	—	23	46	15.2	1.93

(Contd.)

TABLE 10.3 (*Contd.*)

1	2	3	4	5	6	7	8	9	10	11	12	13	14	15
Kulakhet (Rakhtar)	21.5	13.4	6.7	115	95	210	8.4	2.5	50	—	33	50	17.7	1.44
Timrikhet (Mehsasa)	23.5	13.5	6.8	275	85	260	7.8	5.0	45	—	22	56	15.2	4.36
Ukhniyan (Khuna Amal)	20.8	15.2	6.7	225	400	625	7.8	5.0	45	—	32	54	17.6	2.41
Prantkey (Khuna)	20.6	15.6	6.8	325	175	500	7.4	5.0	90	—	42	120	41.7	3.85
Bai-ka-pani (Kandoi)	24.6	13.6	6.8	100	400	500	7.2	2.5	75	—	35	70	21.6	3.87
Baisgar (Kandoi)	20.2	15.1	7.0	150	375	525	7.8	6.0	95	—	37	90	31.3	2.89

Note : All values except pH and temperature are in mgl^{-1} (ppm), unless otherwise mentioned.

TABLE 10.4: Physio-chemical characteristics of surface water in Nagthat

Source (near to village)	*Ambient temp. (°C)*	*Water temp. (°C)*	*pH*	*TDS*	*TSS*	*TS*	*DO*	*FCO²*	*HCO_3^-*	*CO_3^-*	*Chloride*	*TH*	*Ca^{++}*	*Mg^{++}*
Rain water in *khal* (Lohani)	20.8	25.1	5.0	625	500	1125	10.6	33.0	20	—	41.0	56	18.4	2.41
Kafed waterfall (Gadaul)	15.4	9.6	6.4	180	150	230	10.4	2.0	30	—	27.0	36	10.4	2.41
Chipli waterfall (Gadaul)	18.5	9.7	6.5	425	225	650	10.4	2.0	25	—	18.0	20	6.4	0.96
Turiyakhara *nala* (Rampur)	22.6	12.8	6.5	550	110	660	8.2	4.0	40	—	25.0	38	10.4	2.90
Ghasia *nala* (Chichrar)	23.8	14.0	6.6	560	85	645	8.8	2.0	50	—	18.0	46	15.2	1.92

Note : All values except pH and temperature are in mgl^{-1} (ppm), unless otherwise mentioned.

quality varies from place to place depending upon the characteristics of the strata and type of the underlying rocks. The time of contact of water with the bed rock may be different for different locations and this may result in some variations in the water quality of the springs (Drever 1982, Kumar & Rawat 1996).

The pH of the spring waters was observed in the range of 6.3 to 7.0 (average 6.5) while in the surface waters it ranged between 5.0 to 6.6 (average 6.2). Most of the pH values for the spring water and the surface water samples exceed the permissible limit for drinking purposes (ISI 1993). In general pyrite and calcareous minerals present in the system have the greatest impact on the aqueous regime. As a result of low pH there exists danger of heavy metals dissolving easily in the water making it toxic and thus adversely affecting the drinking water quality of these sources (Singh & Choudhary 1997).

It was interesting to record that the water from Kuasa, Rakhtar, Khuna, Chichrar and Mehsasa villages (located in lower altitudes without appreciable forest cover) and the Nagthat springs (at high altitude but without forest cover) have higher pH (6.7 to 7.0) values than the springs located at the higher altitudes with dense forest around. The possible reason for this could be the leaching of the organic matter (litter). Similar results were observed by Kumar and Rawat (1996) while working on the natural springs of Almora. The lowest value of pH (5.0) in the *khal* water may be attributed to direct surface runoff from pyritic minerals and the evaporation.

Spring water samples were observed to be more transparent than the surface water samples as a result of lower total solid contents (TS). The value of total dissolved solids (TDS) varies between 85 and 425 mgl^{-1} in spring water and from 180 to 625 mgl^{-1} in surface water samples, the suspended solids however varied between 85 and 400 mgl^{-1} and 85 to 500 mgl^{-1} respectively in spring and surface waters (Fig. 10.1).

TDS in some of the water samples and TSS in the majority of the samples are found to be beyond the maximum permissible limits set by the Indian Standards

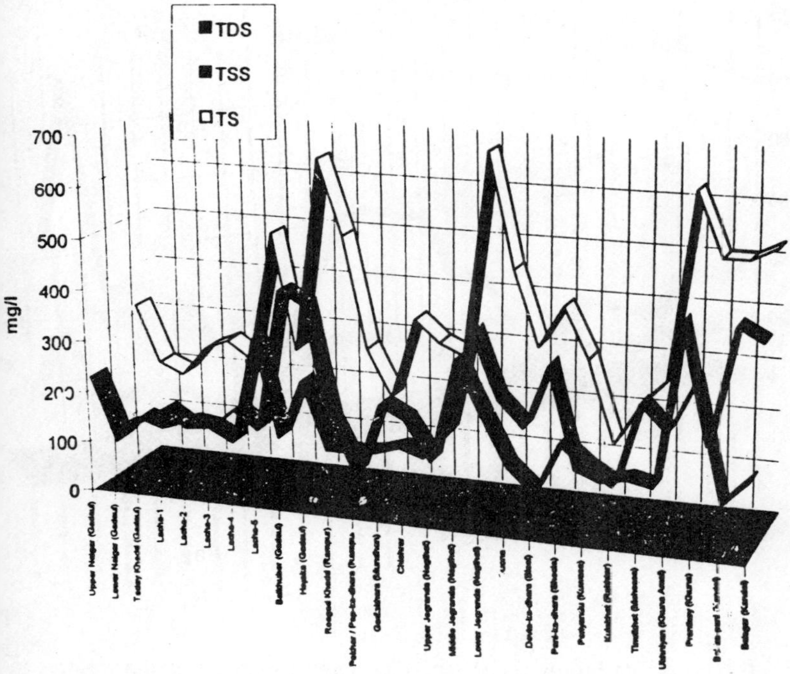

FIG. 10.1: Variation of the solids in the spring waters of Dabred watershed

(1993). Direct mixing of the percolated dissolved substances and comparatively low volume of water might be responsible for the increased concentration of the TDS while the higher values of TSS may be due to the movement of dust particles from open crop fields and the road construction sites.

The values of total (TH) and calcium hardness were higher than the magnesium hardness. Total hardness ranged from 18.0 to 120.0 mgl^{-1} and 20.0 to 56.0 mgl^{-1} in spring and surface water respectively (Fig. 10.2).

There is no permanent hardness recorded in the present study and wide range of fluctuation was observed in both the water sources at different locations. Except for two water sources, i.e. at Khuna and Kandoi villages, the rest

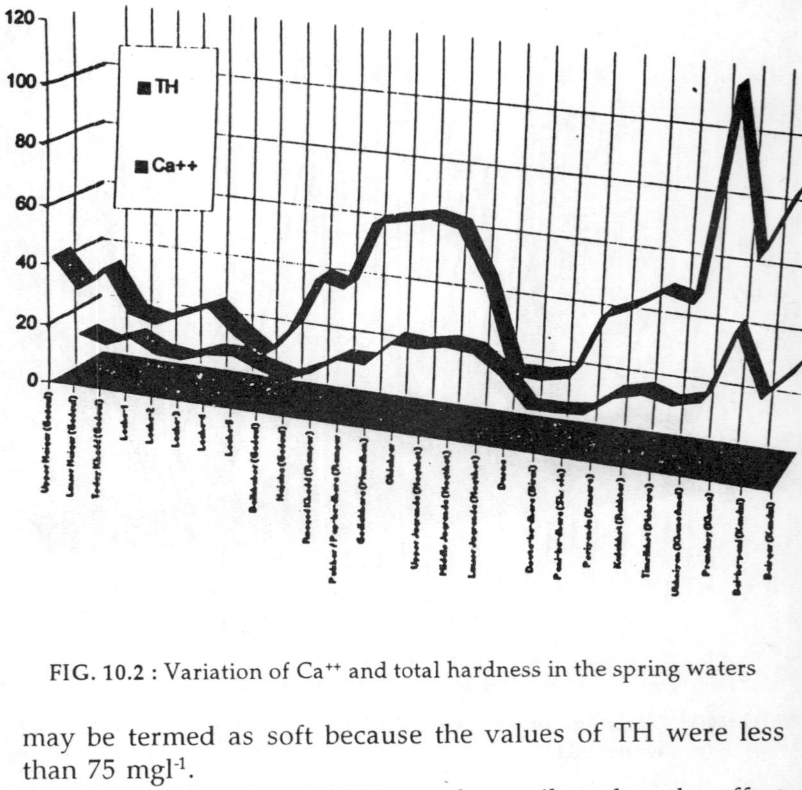

FIG. 10.2 : Variation of Ca^{++} and total hardness in the spring waters

may be termed as soft because the values of TH were less than 75 mgl^{-1}.

The lower values of TH may be attributed to the affect of rainfall that tends to decrease the values of bicarbonates and ultimately the hardness (Jana 1973). Total hardness is essential from potability point of view as there is a danger of heavy metals dissolving easily in soft water with low pH, and making it toxic. All the values of TH were found to be well within the desirable limits as per ISI (1993) specifications for drinking water quality. Both surface and spring water is dominated by Ca^{++}; with low Mg^{++} relative concentration. Mg^{++} concentration ranged between 0.51 to 4.36 mgl^{-1} and 0.96 to 2.90 mgl^{-1} in spring and surface water sources respectively. Both calcium and magnesium in water are desirable from potability point of view. The desirable limit for calcium and, magnesium are 75.0 and 30.0 mgl^{-1} as per the ISI (1993) directives. In the present study all the samples fall within this limit.

Based on these studies it can be stated that the water of the sampled springs is adequately safe and does not pose any serious threat to the health of the masses. Better practices of water conservation however, need to be popularised in the area. Most of the agricultural lands in the area are rainfed and the masses have not derived advantage of the numerous natural water sources they are gifted with. It seemed that the masses are waiting for some external intervention. Social energization is, therefore, required in the region to break the inertia. The spring water of the region if managed properly has the potential of enhancing the yields of the agricultural lands enormously. Both ferrocement and LDPE tanks should be promoted to harvest rainwater that would be available at a short distance besides a network of large storage tanks to avoid wastage of spring water. This water should be utilised through a network of *kuhls*. In all these activities people's participation should be ensured and all major decisions be taken with the concurrence of the majority opinion.

11

DISCUSSION UPON THE CAUSES OF WATER SCARCITY

It is generally accepted that the natural environment of Himalaya is deteriorating. Several theories have been put forth to explain the causes. Ironically, several of the widely accepted assumptions for the deterioration are either without factual support, or are demonstrably unsupportable (Ives & Messerli 1989). Synthesis of literature indicate some common points of different theories:

- Unprecedented population growth followed the introduction of modern health care systems after 1950.
- Immigration from downstream is adding to the growing population of the hills.
- Increased population in the hills has led to:
  - reduced amount of land per family.
  - deepening poverty
  - massive deforestation.

The impact of these factors has been described as the vicious circle, in that the needs of the burgeoning subsistence population are exerting increasing pressures on the forest cover leading to massive deforestation and is the cause of accelerating rates of mass wastage and reduced water availability.

The deforestation, that includes cutting of agricultural terraces on steeper and more marginal mountain slopes, has led to a catastrophe increase in soil erosion, and loss of productive land through accelerated rated of mass wastage, and to the disruption of the normal hydrological cycle. This situation, has in turn, led to increased runoff during the monsoon together with disasterous flooding and massive siltation in the plains, and lower water levels and drying up of springs and wells during dry season. Related ills are: rapid siltation of reservoirs; abrupt changes in the courses of rivers; spread of barren sand and gravel across rich agricultural land on the plains; and increased incidence of diseases in the downstream areas. Deforestation is also perceived as the sole cause for reduced rainfall.

Facts

An average female in rural areas of Himalaya work for long hours daily; her day starts in the dawn and is at work till late in the night. Besides the usual household chores she has to look after almost all agricultural activities and ensure fodder, fuelwood and water supplies.

The female generally brings 15 litres of water at one time and on an average it takes nearly 45 minutes to fetch water from the nearby source, but the actual process sometimes takes as long as five hours. This time can be gainfully utilised for other income generating activities.

Tired of fetching water, the females try to use as little water as possible for household purposes. According to one survey the households where the females have to bring water use just 13 litre of water daily in contrast to 30 to 100 litre used by those who have the privilege of having a tap connection. This leads to unhygenic living conditions and may cause serious health hazards.

Of all the resources, water is the most important. By virtue of its dynamic nature it affects the distribution and quality of other resources. Water thus forms an important element of the theory of Himalayan degradation discussed above. The scientific management of this resource has the potential of improving the quality of life of the masses to a great extent. Water resources' development or management represents an attempt to meet existing or potential demand for water. Attempts to meet these demands are commonly based upon engineering modifications of the existing water supplies. These in turn, are dependent upon a quantitative understanding of the temporal and spatial variability of the supplies. It is apparent that a major component of attempts to develop or manage water resources in any environment involves scientific studies of the resource, in advance of need, if project planning is to be realistic. Within any river basin, attempts to meet demand that exceed the natural supply of water will be, at best, only partially successful.

There are thus two primary needs in understanding the water resources of any region: it is essential to assess accurately the existing demands for water, and the nature of the water supply must be understood if these demands are to be met. Demographics determine the nature of demand upon the water resources. Economics and engineering determine the extent of technologies available and the willingness of the society to apply those technologies. Hydrology determines the extent to which the supply will meet the demand.

In the context of factors affecting the flow of water through a watershed, the environment may be defined as:

- Climate
- Geology
- Topography or terrain, and
- Vegetation

Environmental management from a water resources' perspective, is a matter of determining the extent to which a change in any one, or some combination, of these factors will produce a corresponding change in the timing, volume, or quality of water flowing through the watershed. Alternatively, it is a matter of inducing a change in one or more of the factors so as to produce some desired change or correct a perceived imbalance. Environmental interactions are complex, and at least two of the factors—climate and geology—can not be managed in any meaningful way. In the case of these two factors, it is essential to understand how each relates to the water and sediment balance in the watershed, but the only realistic management alternative is to make allowances for the prevailing conditions of climate and geology in the management planning.

Terrain and vegetation can be managed, in the sense that the type and density of vegetation cover may be purposefully altered and the terrain may be shaped, or altered by a variety of engineering structures. The extent to which changes in terrain or vegetation will affect the water or sediment balance will depend on the existing conditions of geology and climate, the nature of changes in vegetation or terrain, and the scale on which the change is effected. In all cases, it must, however, be recognised that those factors that may be managed may not be critical factors in determining the water and sediment balance. Only a study of the interrelationships at a specific site can determine if environmental management of any sort will have a positive or negative effect on the water resources at, and downstream from, that site (Hamilton 1987).

Water is an integral aspect of all facets of the mountain environment and enters into virtually every resource development or management project in some way. All disciplines involved in resource planning or management have developed numerous technologies for modifying the natural characteristics of volume, rate, timing, or quality of water at a site. Basically, all technologies are variations of a few basic themes: storage of water during a period of surplus for use in a period of deficit, transfer of water from an area of surplus to one of deficit, and improving water quality. While there are only a handful of ways in which the hydrological regime of an area may be changed beneficially, in practice there are virtually endless number of variations, combinations, and permutations of these basic themes (Dunne & Leopold 1978).

In the mountainous regions the most critical factors are topography and meteorology (Barry 1981, Baker 1944). The topographic factors of local relief, slope angle, and aspect influnce the timing, volume, and spatial variability of water and energy in the mountainous terrain (Alford 1985). In the mountains there is no regional climate as such, but rather a mosaic of local topoclimates (Thornthwaite 1953) determined by variations in slope angle, aspect, and relative altitude (Flohn 1974, Geiger 1966) and the interaction between topography and meteorology produces a situation in which the following takes place:

Precipitation varies complexly with the aspects of altitude and terrain. There is commonly an orographic gradient, in which precipitation amounts vary along altitudinal gradients. Generally, windward slopes are wetter than leeward slopes. Evaporation losses decrease with altitude as available energy decreases (Lambert & Chitrakar 1987).

Steep mountain slopes cause water produced by rain or snowmelt on the surface to run off quickly into stream channels (Petts & Foster 1985). In many cases, shallow mountain soils and impermeable geologic formations can provide little storage for soil moisture and groundwater. Vegetation may be zoned based on both altitude and aspect, limiting the hydrological impact of either removal or replacement to narrow geographical limits for any single mountain watershed.

Erosion includes all processes that result in physical lowering of the surface of the earth. Surface erosion is the loss of surface materials, as a result of the action of falling and running water, or wind, while mass wasting involves the movement of large masses of fractured bedrock or other unconsolidated materials, including soils, from a slope. Sediment transport is the movement of the products of erosion through a river system (Leopold *et al.* 1964). It is not possible to consider water resources' management in the mountainous terrain without considering the sources of very large volumes of sediment moving annually through the river systems of the region. The values of the sediment load for this region exceed the world average by almost two orders of magnitude.

Fractures and other void spaces within the geologic materials are the primary factors in determining the existence and importance of groundwater resources. Where the rocks neither porous nor permeable at depth there is no groundwater, no matter how wet the surface environment may be. The best groundwater environments are well-sorted alluvium, cavernous limestones, or sandstones.

Mountain watersheds are poor groundwater environments. The rocks forming the core of a mountain range generally have limited porosity and permeability, limiting groundwater occurrence to local fractured zones or pockets of alluvium. Significant groundwater resources most often begin only at margin of the mountains—the piedmont zone—and extend downwards into the alluvial valleys of the lowlands (Hagen 1980).

It becomes clear from the foregoing discussion that though water constitutes an important parameter affecting the environmental degradation; its interaction with others is complex and often hard to understand and control. The entire Himalayan region is experiencing acute shortage of this important resource. Many reasons have been put forth to explain the depleting water resources of the region. Some of these are:

- Increasing population;
- Manoeuvres in the recharge zone of the springs;
- Engineering activities;
- Deforestation;
- Vanishing pasture lands;
- Recurring forest fires;
- Increasing pollution;
- Change in attitude and life style;
- Change in cropping pattern;
- Increasing industrialisation;
- Neglect of traditional sources;
- Disruption of the social mechanism of resource conservation; and
- Change in the hydrology of the watersheds by tapping of water at higher elevations.

A thoughtful analysis would reveal that the shortage of water is directly linked to the vicious circle of resource degradation. Increasing population onsets deforestation for agricultural and allied activities. This reduces recharge of groundwater and thus the spring yields at the face of increasing demand. To fulfill this increasing demand more sources are tapped on the higher reaches. This leads to depleting yields of the downslope springs and thus shortage of water in these areas. Consummerism leads to the introduction of high yielding, water demanding crops at the cost of traditional hardy crops and the social mechanisms for the conservation of resources. Water requirements of these crops are hard to meet and this leads to crop failure and the ultimate impoverishment of the society.

We are living in a threshold period and if we do not plan and act early the region would be pushed to the verge of an eventual drought. Drought, aridity and ultimate desertification—this is what we are leading Himalayas to. The signs of desertification can clearly be observed in nature where many exotic and unheard of xerophytic plant varieties are fast displacing the native plant species. The plant species are perhaps reacting to the anthropogenically induced stress. The plant community is certainly behaving more intelligently than we humans and preparing itself for the eventual onslaught. Its not the matter of money and had it been so the things would have been far better. What is required today is sincere and multipronged, multidisciplinary effort. What is needed is participation—participation of the masses, participation of the people at grassroot level. So it is the sincere effort from within ourselves that alone is capable of averting the catastrophe writ large in the destiny.

Water and Forests

The most emotional positions taken concerning the relationship between an environmental element and water resources has centered around the role played by vegetation in the hydrological cycle (Bowonder 1982, Haigh 1989, WRI 1985). This is particularly true of the relationship between changes in forest cover and possible changes in the timing or volume of water availability, both within and downstream from the watershed, as well as erosion within and sediment transport from the affected watershed. Practical experiments in this regard many a times contradict well established beliefs. It is without any bias different views regarding the role of forests are put-forth. Vegetation plays three primary roles in the hydrologic cycle: (i) it acts as a buffer between the falling or flowing water and the ground surface, (ii) it intercepts a percentage of precipitation which is then returned directly to the atmosphere by evaporation, and (iii) it returns to the atmosphere a portion of the water which falls as precipitation by the process of transpiration—the flow of water through the roots, stems, and leaves of plants that accompanies photosynthesis.

Vegetation and surface water and energy environment evolve together. This fact is reflected in concept of plant succession within ecosystems. In the absence of disturbance, plant communities follow one another in somewhat predictable sequence and the surface environment is altered by each in turn. In determining possible impacts of vegetation multiplication, either removal or replacement, it is useful to understand that, at least from the perspective of hydrology, a forest is much more than a collection of trees. A forest is a complex assemblage of trees, bushes and other understory plants, ground litter, soil, and often in the mountains, bedrock. The assemblage of environmental elements modify the impact of rain drops on the surface, increase the water infiltration potential and alter the nature of energy exchange processes on or near the surface. It is the assemblage of elements, and not the trees alone that modulate the hydrological cycle. Removal of the trees, if accompanied by a loss of understory vegetation, litter and soil as well, may well produce measurable changes in the flow of water and energy, into, through and from the site. Replacement of trees—reforestation—will not similarly restore the pre-existing conditions until sufficient time has passed for the elements of the mature forest—the understory, litter, and soil—to have been replaced.

It is not necessarily the type of vegetation that is important but rather the density of the ground cover that is critical in determining rates of erosion. Changes in the water and sediment budgets of watersheds can be expected as a result of changes in the type or density of the vegetal cover (Dunne & Leopold 1978).

In the Himalayas the Chir Pine monoculture is viewed by the local people as an attack on their life support system

(Shiva *et al.* 1981). A widespread misunderstanding is the oft-repeated assertion that tree roots are a sponge which soak up rainwater during wet seasons and release it slowly during dry seasons—that is, that the forests act as hydrological regulators (Spears 1982; Myers 1983). The implications are that cutting down forests causes floods during the summer monsoon season or during major storms, and results in lower water levels, or in rivers totally drying up during the dry seasons (Spears 1982; Myers 1983). Watershed research however suggests that tree roots behave more like a pump than a sponge. Forest cutting on 94 small catchments world wide, compared with uncut control catchments, resulted in more total water yield from the cut-cover areas, with the increase being proportional to the intensity of the cutting (Bosch & Hewlett 1982). This occurs because tall forests evaporate and transpire more water than other types of vegetation. Furthermore, Hamilton's (1983) review of catchment research showed that in most cases following cutting, increased flows occurred throughout the year, including the dry season. Gilmour (1977) demonstrated that a stream that previously had ceased to flow during many dry periods flowed perennially after logging. Hamilton (1987) points out a notable exception to this effect in the case of cloud forests that intercept precipitation from wind driven clouds.

The line of reasoning can be applied to the alleged role of the upper forest canopy in protecting the soil from raindrop impact; lowering of groundwater table; water yield; and timing and distribution of stream flow. With the intact upper forest canopy rainwater accumulates on the leaves and fall on earth as much bigger drops than raindrops. Also, given a height in excess of 20 m the terminal velocity of the larger water drops will not be significantly different from that of raindrops falling on the ground unimpaired. Furthermore, it is the litter and the low vegetation that protects the soil from particle detachment by raindrop impact. Fire or grazing, that remove low vegetation, and litter gathering that exposes bare soil should be reduced, rather than worrying about reduction in the tree canopy if the principal concern is sheet erosion.

With the surge in productivity of a shrub and herb layer following forest clearing with more solar radiation reaching ground level, the soil surface may be actually better protected against raindrop impact.

Cutting of forests, and also their replacement with a different landuse have usually resulted in the free water table moving closer to the surface, because less water is lost to evapotranspiration (Hamilton 1987). Tree cutting alone does not lead to increases in surface erosion—what is important is the mode of forest harvesting and ground cover.

Much of the sediment in the given river basin is in various temporary storages; thus claims of immediate sediment reduction on a large scale and with far distant effects following any afforestation project should be critically reexamined.

Many blame deforestation for the catastrophic flooding (Reuters 1985). The real cause is too much rain occuring in too short a time. Some areas sometime receive exceptionally high rains during monsoon season when soils are already saturated. Even if the entire basin is well forested the chances of floods are equally high. Forests are beautifully functioning ecosystems, capable of producing many products and services on the sustained basis. These must in places be totally protected, and in other places wisely managed. These can reduce the magnitude but will not prevent floods or sedimentation in the lower reaches of major rivers,' nor significantly reduce flooding in major storm events (Hamilton 1987). Whatever be the real cause of reduced discharge of the springs it is a fact that the springs of Uttarakhand are showing depletion in yields and the masses are faced with the consequences.

Water Resources : Some Suggestions

— Equitable distribution of water resources amongst urban and rural population;
— Penalty for those polluting the water resources;
— Encouraging effective water use;
— Encouraging water recycling plants;

- — Minimising beaurocratic interference in rural water supply schemes;
- — Ensuring free and easy access to water resources;
- — Improvement in household gadgets so as to bring about economy in water use;
- — Effective and well managed water supply system;
- — Promoting water economy through awareness drives; and
- — Promoting low volume flush systems in latrines.

All water that we use is the outcome of the atmospheric precipitation, in one form or the other. Precipitation in a given geographical area is a function of a vast number of variables and is bound to fluctuate within given limits. Our needs are, however, not bound by any limits and are showing manyfold increase. So very soon all places are going to face shortage of water. So not going into the debate any further it would be important to discuss some measures for improving the situation.

If the drinking water requirements of the people habitating the remote hilly areas are to be met in a cost effective and sustainable way, in a manner that is ecofriendly and is aimed at reducing the dependence of the local people on external inputs, a sound strategy aimed at reviving the drying springs and enhancing the yield of the existing springs has to be made. Such a strategy when implemented successfully, shall certainly be reducing if not solving the water related problems of this region. Moreover, these interferences would impart a long term solution and would be making our small communities self reliant by reducing dependence upon external inputs and recurring cost of maintenance would be cut short. Moreover, the people habitating the area would reap many indirect benefits that would be reflected in their improved economy and quality of life.

It becomes important to understand the inter-relationship between spring yield and the recharge. Spring yield in any area is a function of recharge that is dependent upon the time duration for which the water stays at higher elevations. The longer this period, the higher would be the recharge. The vegetal cover particularly the undergrowth, grasses and the litter impede the flow of water downslope. The roots provide the channels for the water to percolate down.

There are many ways to enhance recharge. Some of these are:

- Contour aligned canals—canals should be contour aligned and clay/mud lined so as to keep water at higher elevations and to permit large recharge;
- Clay lined percolation/water harvesting tanks at the slope break;
- Recharge pits at higher reaches;
- Retention walls in gully slopes together with vegetal impedements;
- Deep and contour aligned ploughing in the fields;
- Maintenance of *bundhs* of the fields;
- Terracing of steep slopes;
- Plantation of suitable species;
- Development of pasture lands; and
- Recharge pits for the household waste water.

Entire Uttarakhand region has a long and outstanding tradition of water harvesting and many examples of water harvesting practices together with socio-religious practices for the maintenance of these and the recharge zone are observed in the region. With the onslaught of tapped water these have however been sidelined. Traditional water harvesting structures should be woven into the organised water distribution system. Proper maintenance of these structures would help in tiding over crisis besides reducing the pressure upon the existing infrastructure.

It is important to appreciate the potential of rainwater harvesting. A series of studies carried out by different agencies and also by the author (and discussed in the previous chapters) have shown that the water precipitating in any watershed area is sufficient to meet the drinking water needs of the local population. Moreover, the technology involved is simple and cost effective, and can be put to use easily in the hilly houses with sloping roofs.

These studies show why our ancestors found water-harvesting structures so important and useful so as to include them into the regio-cultural fabric of the region. Trees for Life (Ranichauri) is taking water harvesting in a serious manner and promoting it in the remote villages of Tehri Garhwal with active participation of the villagers. A series of Low Density Poly Ethylene (LDPE) and ferrocement tanks have been erected in the villages to collect the rainwater together with *khals/chaals* and other water harvesting structures at the upper reaches. These humble efforts have the potential of making the Himalayan communities self-reliant in water.

Lengthy pipelines have proved out to be unsuccessful in this fragile and hazard prone zone because of its unimaginably high cost of maintenance and energy. Traditional water sources, therefore, need to be revived and it is these that are capable of giving solace to the masses of this region. Only local springs and small streams in the area have the potential of making our community self-reliant. Better recharge zone management practices need to be popularised in the region with the active involvement of the masses. Appeal through the little tradition of the region in this regard can be highly effective.

The proper and scientific management of water will be reflected in :

— Reduced rates of mass wastage;
— Increased fertility and thus crop yields;
— Easy access to energy; and
— Enhanced fodder and fuelwood availability.

Till date prospects of groundwater exploitation have not been studied seriously in the hilly terrain. With the help of satellite remote sensing and geophysical surveys potential groundwater zones should be identified in the region.

Besides emphasising economy in the use of this important resource, which by virtue of its cheap availability is often not considered a resource, modern water conserving agronomic practices need popularisation together with recycling of waste water.

Irrigation is mostly rendered free or at highly subsidised rates. So is the case with drinking water. This has triggered abuse of this scarce resource that is getting scarcer with each day that passes by. Supply of water should be metered and adequate taxes should be imposed upon the water supplied; at least on the well-off sections of the society, commercial and governmental establishments and large farmers. This measure would help in stopping the rampant abuse of this scare resource.

NGO movement should be strengthened and people's participation should be ensured in the state-sponsored and other schemes.

BIBLIOGRAPHY

Agarwal, A. & Narain, S. (1997), *Dying Wisdom: Rise, fall and potential of India's traditional water harvesting systems*, Centre for Science & Environment, New Delhi.

Alford, D. (1985), Mountain hydrologic systems, *Mountain Research & Development*, 5, 349-63.

Alford, D. (1992), *Hydrological aspects of the Himalayan region*, ICIMOD Occasional Paper, 18, 68p.

APHA (1989), Standard methods for the examination of water and waste water, 17th Edition, *American Public Health Association*, Washington DC.

Baker, F. (1944), Mountain climates of western United States, *Ecological Monographs*, 14, 223-54.

Bandhopadhyay, J. (1987), Political ecology of drought and water scarcity—Need for an ecological water resource policy, *Environmental Protection Works*, XXII (80), 2159-69.

Banerjee, J.M. (1967), *History of Feroz Shah Tuglaq*, Delhi.

Barry, R. (1981), *Mountain Weather and Climate*, Methuen & Co. Ltd. London.

Bartarya, S.K. (1995), Hydrology and water resources of intermontane Doon valley, *Journal of Himalayan Geology*, 6, 1995, 17-28.

Bosch, J.M. & Hewlett, J.D. (1982), A review of catchment experiments to determine the effect of vegetation changes on water yield and evaporation, *Journal of Hydrology*, 55, 3-23.

Bowonder, B. (1982), Deforestation in India, *International Journal of Environmental Studies*, 18, 223-36.

Central Water Commission (1991), *Water and Related Statistics*.

Chhabra, B. & Gai, G.S. (1981), Inscriptions of the early Gupta kings, In: *Corpus Inscriptionum Indicarum*,

Archeological Survey of India, New Delhi, 3, 296-305.

Dewan, M.L. (1990), *People's Participation in Ecosystem Development*, Concept Publishing Co., New Delhi, 274 pp.

Dhyani, B.L., Ram Babu, Sewa Ram, Katiyar, V.S., Arora, Y.K., Juyal, G.P., & Vishwanatham, M.K. (1993), Economic analysis of watershed management programme in outer Himalayas—A case study of ORP Fakot, *Indian Journal of Agriculture and Economy*, 48, 237-45.

Dowson, J. (1953), *Tarikh-e-Feroz Shah of Shams-I-Siraj Afif, The history of India told by its own historians*, Calcutta.

Drever, J.I. (1982), *The Geochemistry of Natural Water*, Prentice Hall Inc, London.

Dunne, T. & Leopold, L. (1978), *Water in Environmental Planning*, W.H. Freeman & Co., New York.

Flohn, H. (1974), Contributions to the comparative meteorology of mountain areas, In: *Arctic and Alpine Environments*, J. Ives & R. Barry (eds.), Methuen London.

Geiger, R. (1966), *The Climate Near the Ground*, Harvard University Press, Massachusetts Cambridge.

Gilmour, D.A. (1977), Effect of logging and clearing on water quality in a high rainfall zone of northeast Queensland, In: *Hydrology of Australia*, Institution of Engineers, Australia, National Conference Publication, 77.

Goel, P.K. (1980), Impacts of sewage, *International Journal of Economics and Environmental Sciences*, 6, 86-96.

Goel, S.L. & Jain, R.K. (1987), Administration of environmental health, In: *Environmental Pollution and Health Hazards in India*, Kumar, R. (ed.), Ashish Publishing House, New Delhi, 29-52.

Gupta, R.K. & Katiyar, V.S. (1982), Rainwater cistern systems for the Himalayan region, *Proceedings International Conference on Rainwater Cistern Systems*, Honolulu, Hawaii, USA.

Habib, I. (1969), Potentialities of capitalistic development in the economy of Mughal India, *The Journal of Economic History*, 29.

Hagen, T. (1980), *Nepal*, Oxford and IBH Publishing Co.,

New Delhi.

Haigh, M. (1989), Deforestation and disaster in northern India, In: *Studies in Himalayan Ecology*, T.V. Singh & J. Kaur (eds.), Himalayan Books New Delhi, 67-80.

Hamilton, L.S. (1983), *Forest and Watershed Development and Conservation in Asia and the Pacific*, Westview Press, Boulder Colorado.

Hamilton, L.S. (1987), What are the impacts of Himalayan deforestation on the Ganga—Brahmaputra lowlands and the delta? Assumptions and facts, *Mountain Research & Development*, 7, 256-63.

Hengeveld, H. & De Vocht, C. (1982), *Role of Water in Urban Ecology*, Elsevier Scientific Publishing Company, Amsterdam.

Hussenappa, V., Sastry, G. & Khullar, A.K. (1980), Rainfall analysis for water harvesting in Doon valley, National Symposium on Soil Conservation and Watershed Management, CSWCRTI Dehradun.

ISI (1993), Indian standard specification for drinking water, IS: 10500-1993, *Indian Standard Institute*, New Delhi, 6-11.

Ives, J.D. & Messerli, B. (1989), *The Himalayan Dilemma*, Routledge London & New York, 295p.

Jana, B.B. (1973), Seasonal periodicity of plankton in fresh water pond in West Bengal, *Indian Inst. Rev. Ges. Hydrobiol.*, 58, 127-43.

Jha, Pt. A. (1988), *Vrhat Sanhita (550 AD) by Varahamihira*, Chow Khamba Vidyabhawan, Varanasi.

Juyal, G.P. & Gupta, R.K. (1985), Construction of LDPE lined tankas in hills—A case study, *Indian Journal of Soil Conservation*, 13, 10-13.

Juyal, G.P. & Katiyar, V.S. (1991), Water resource development and management in small hilly watersheds, *Journal Indian Water Resources Society*, 11, 14-17.

Kangle, R.P. (1963), *The Kautilya Arthasastra*, University of Mumbai, Part I & II.

Krishnamurthi, B.V. & Schoettli, U.R.S., Environment in India's religious and cultural heritage, In: *India's Environment: Crisis and Responses*, Bandhopadhyay, J.,

Jayal, N.D., Schoettli, U. & Singh, C. (eds.), Natraj Publishers Dehradun, 159-71.

Kumar, K. & Rawat, D.S. (1996), *Water Management in Himalayan Ecosystem: A Study of Natural Springs of Almora*, Indus Publishing Company, New Delhi.

Kumarappa, J.C. (1930), *Public Finance and our Poverty*, Navajivan Publishing House, Ahmedabad.

Lambert, L. & Chitrakar, B. (1987), Variation of potential evapotranspiration with elevation in Nepal, *Mountain Research & Development*, 9, 145-52.

Leopold, L., Holman, M. & Miller, J. (1964), *Fluvial Processes in Geomorphology*, W.H.Freeman San Francisco.

Metcalf & Eddy (1979), *Wastewater Engineering; Treatment, Disposal, Reuse*, Tata McGraw Hill, New Delhi.

Myers, N. (1983), Tropical moist forests: over exploited and under utilised,?, *Forest Ecology & Management*, 6, 59-79.

Naudou, J. (1971), An example of engineering in Medieval India, In: *Prof. K.A. Nilakanta Sastri Felicitation Volume*, Madras, 224-30.

Nehru, J.L. (1946), *Discovery of India*, Reprint 1989, Oxford University Press, New Delhi.

Pande, R.K., Pande, Y.C. & Gangola, G. (1992), *Quality of Urban Water Supply in Central Himalaya*, Conserving Wildlife in India, 89-101.

Pant, N. (1989), Farmer managed irrigation systems n India, FMIS Network, *International Workshop on Design Issues in Farmer Managed Irrigation Systems*, Chiang Mai, Thailand.

Pant, P.C. (1995), *Parvatiya Jal Strot* (in Hindi), Shree Almora Book Dept, Almora.

Petts, G. & Foster, I. (1985), *Rivers and Landscape*, Edward Arnold Publishers Ltd. London.

Possehl, G.L., & Raval, M.H. (1989), *Harappan Civilization and Rejdi*, Oxford IBH, New Delhi.

Rao, E.G.K. & Kotekar, P.O. (1971), Exploration of underground water springs according to the ancient Hindus, *Indian Journal of History of Science*, 6, 139-46.

Ratnagar, S. (1986), An aspect of Harappan agricultural production, *Studies in History*, 2, 137-53.

Rautela, Meeta (1996), Ecobiology of a snow trout *Oreinus*

richardsonii (Gray) from Sarju river, Kumaun Himalaya, Unpublished Ph.D. thesis, Kumaun University, Nainital.

Roy, A.K. (1995), Water resources in Doon valley: Development and constrains related to hydrological environment, *Journal of Himalayan Geology*, 6, 123-25.

Sawyer, C.N. (1960), *Chemistry for Sanitory Engineers*, McGraw Hill, New York.

Sharma, J.S., Sharda, V.N. & Sikka, A.K. (1996), *Water Harvesting and Recycling: Indian Experiences*, Central Soil & Water Conservation Research & Training Institute, Dehradun, 248p.

Sharma, Y.D. (1974), *Delhi and its Neighbourhood*, New Delhi.

Shiva, V., Sharatchandra, H.C. & Bandyopadhyay, J. (1981), Social, economic and ecological impact of social forestry in Kolar, Indian Institute of Management, Bangalore, India (mimeo).

Singh, R.B.P. & Choudhary, S.K. (1996), Trace metal in surface and ground water in Simlong coal mining area, Rajmahal, Bihar, *Proc.Nat.Sym.on Plant Diversification and Human Welfare*, 5 & 6 April 1995, Bhagalpur University.

Singh, R.B.P. (1997), Impact of coal mining on water, soil and vegetation of Simlong area, Unpublished Ph.D. thesis, Bhagalpur University.

Spears, J. (1982), Rehabilitating watersheds, *Finance & Development*, 19, 30-33.

Srinivasan, T.M. (1970), A brief account of the ancient irrigation engineering systems prevalent in South India, *Indian Journal of History of Science*, 5, 315-25.

Thadhani, P.K. (1985), *Chronicles of the Doon Valley: An Environmental Expose*, Indus Publishing Company, New Delhi.

Thomas, C.P. & Backford, S. (1982), Tracing well water pollution in a limestone aquifer, *Journal of American Water Works Association*, 74, 182-87.

Thornthwaite, C.W. (1953), *Topoclimatology*, Symposium on Microclimatology & Micrometeorology, Toronto Meteorological Conference, Toronto, Canada.

Tiwari, R.S. & Garg, P.K. (1995), Landform and landuse of

Doon valley and their role in resource development, *Journal of Himalayan Geology*, 6, 9-15.

Welch, P.S. (1948), *Limonological Methods*, McGraw Hill, New York.

WHO (1985), Guidelines for drinking water quality, *Recommendations*, WHO, Geneva.

World Resources Institute (1985), *Tropical Forests: A call for action*, WRI Washington D.C.

INDEX